Pembrokeshire College
Learning Reso

D1421010

BROKESHIRE COLLEGE

R07561X0618

Before and After
an Oil Spill

Before and After
an Oil Spill:
The Arthur Kill

edited by Joanna Burger

Rutgers University Press
New Brunswick, New Jersey

Library of Congress Cataloging-in-Publication Data
Before and after an oil spill : the Arthur Kill / edited by Joanna
    Burger.
        p.   cm.
    Includes bibliographical references and index.
    ISBN 0-8135-2095-9
    1. Oil spills and wildlife—Arthur Kill (N.J. and N.Y.)   2. Oil
spills—Environmental aspects—Arthur Kill (N.J. and N.Y.)   3. Oil
pollution of rivers, harbors, etc.—Arthur Kill (N.J. and N.Y.)
4. Exxon Corporation.   I. Burger, Joanna.
QH545.05B44 1994
363.73′82′0916346—dc20   93-43804
                                                                  CIP

British Cataloging-in-Publication information available

This collection copyright © 1994 by Rutgers, The State University
Individual chapters copyright © 1994 in the names of their authors
All rights reserved
Manufactured in the United States of America

For my parents,
E. Melvin Burger and Jannette Vivienne Male Burger

My husband, Michael Gochfeld

And to my hopes for a sustainable biosphere

Contents

Preface   ix

Contributors   xiii

1   Introduction   *Joanna Burger*   1

Responses and Cleanup

2   Governmental Cooperation
    Paul M. Hauge and Robert K. Tucker   23

3   Legal Considerations   *Gordon J. Johnson*   44

4   The Role of Conservation Organizations
    Carolyn Summers   64

5   Rehabilitation of Contaminated Wildlife   *Lynne Frink*   82

6   Bioremediation and the Arthur Kill   *Joanna Burger*   99

Biological Effects

7   Immediate Effects of Oil Spills on Organisms in the Arthur
    Kill   *Joanna Burger*   115

8   Effects of Oil on Vegetation   *Joanna Burger*   130

9   The Effects of Oil Spills on Bivalve Mollusks and Blue Crabs
    Keith R. Cooper and Angela Cristini   142

10   Fiddler Crabs (*Uca* spp.) as Bioindicators for Oil Spills
    Joanna Burger, John N. Brzorad, and Michael Gochfeld   160

11   Fish and Shrimp Populations in the Arthur Kill
    John N. Brzorad and Joanna Burger   178

12   Gull and Waterfowl Populations in the Arthur Kill
    Alan D. Maccarone and John N. Brzorad   201

13   The Arthur Kill Oil Spills: Biological Effects in Birds
    Katharine C. Parsons   215

viii   Contents

14   Biological Effects: Marine Mammals and Sea Turtles
     Romona Haebler   238

15   The Arthur Kill, People, and Oil Spills   *Joanna Burger*   253

16   Ecological Risk, Risk Perception, and Harm: Lessons from the
     Arthur Kill   *Michael Gochfeld and Joanna Burger*   265

Conclusions

17   From the Past to the Future: Conclusions from the Arthur
     Kill   *Joanna Burger*   283

18   Epilogue: A Matter of Viewpoint   *Joanna Burger*   290

     Index   295

Preface

Wide expanses of swaying *Spartina* fringe the channel, and narrow mud flats glisten in the early morning light. Tiny burrows dot the dark m id, and hundreds of fiddler crabs move slowly over the surface, feeding on a thin film of green algae. Here and there male fiddler crabs wave their large claw to attract females or to repel males looking for territories. Another mud bank is devoid of fiddler crabs. A silent, stalking night heron gazes at the empty holes, waiting for a fiddler crab to emerge. Slowly I drift to the shore of an island at the edge of the channel.

Gentle waves caress the rock-and-sand shores, which are strewn with logs, twigs, and marsh grass. Beyond the wrack, tall green *Phragmites* sway in the breezes, forming a dense curtain. Parting the reeds I walk through the tangle. The reeds give way to small cherry and poplar trees, and then to taller trees. Here the ground is nearly bare, for the canopy above is dense, allowing little light to penetrate to the floor below. Low squawks and moans draw my attention to the thick stick nests lodged in the branches overhead. A gangly, greenish black-crowned night heron chick with a tuft of dark gray-black feathers fringing his forehead peers down at me, then scrambles up and into the dark branches. Nearby, a stately, elegant great egret chick watches my progress. The egret nest is higher in the trees, more difficult to reach then the heron nest, so the egret remains motionless. Only the sounds of the young herons and egrets can be heard as I gently move through the heronry.

Here is a natural ecosystem, and a productive heronry. Yet I pause and consider where I am. This is Pralls Island, in the middle of the Arthur Kill. Part of the New York Harbor and kills complex, the Arthur Kill is one of the most heavily trafficked rivers in the world. It has a high concentration of oil refineries, oil storage facilities, and other industry along its banks. And yet relatively natural ecosystems exist here, in spite of the fact that the Arthur Kill has been exposed to low levels of oil spills and leaks for many years, placing continuous stress on the resident organisms.

When the massive Exxon oil spill during the night of January 1–2, 1990, occurred, it drew media attention to the Arthur Kill, and people began to evaluate the ecological, sociological, and economic importance of the Kill to local residents and the public at large. Various interest groups, governments, and scientists responded to the event, evaluating the damage.

Since the Kill is short and its ecosystem small, it provides a unique opportunity to evaluate the effects of an oil spill. The swift legal settlement allows the events that occurred following the Exxon oil spill to be examined within scientific, conservationist, legal, sociological, and political frameworks. In this volume the key people involved in the damage assessment and legal settlement write from their own viewpoints. Scientists examine the effect of the spill on the organisms of the Arthur Kill, conservationists address their concerns, and government personnel explain their role in the aftermath of the spill.

Although environmental catastrophes often degenerate into an adversarial relationship among the principals, the oil spill in the Arthur Kill generated cooperation and collaboration among many people and organizations. There were, however, disagreements and real differences of opinion over the events and consequences. This book reflects the immediate response and process of damage assessment and legal settlement. The people involved in the response assessment and settlement, including the scientists from universities and research centers, conservationists, rehabilitators, toxicologists, state officials, and attorneys, have written chapters.

In recognition of their valuable data and important perspective, Exxon scientists were invited to write chapters for the book. Several scientists from Exxon began writing chapters but were restricted from publication by their corporate administration. Months of negotiations with Exxon officials almost succeeded in achieving the participation of these scientists, but in the end they were forced to withdraw. Unresolved issues included the writing of rebuttal chapters to each non-Exxon scientist's chapter, excluding all chapters except those from scientists (i.e. legal considerations, conservation perspectives, human use, and risk assessment), and dropping discussion of "existence" or "option" values associated with loss of natural resources.

The data gathered by the able scientists from Exxon has not been lost. As part of the regulatory and legal process Exxon submitted scientific data and reports to the state agencies, and I have used these reports as the basis for the chapters on vegetation and bioremediation. It is my only disappointment with the book that Exxon failed to recognize the importance of its participation in a full examination of the Arthur Kill oil spill.

The overall aim of this book is to chronicle the events of the spill, examine the effects of the spill on organisms (including people) in the Arthur Kill, and provide useful information and insights that will help others who face oil spills or other environmental catastrophes in the future. We have endeavored to examine the Arthur Kill spill from a balanced perspective; thus, all chapters were reviewed extensively by people with different viewpoints. In the end, however, each chapter represents the views and science of its authors. Differences in perspectives can provide insights that are valuable for future oil spills, for only by working together and acknowledging differences can we move ahead to prevent and minimize oil spills in the future and know how to respond to restore ecosystems.

Acknowledgments

Over the years, many people have discussed the effects of environmental insults, including oil spills, with me, and I am grateful for all their insights. I particularly thank Michael Gochfeld, Carl Safina, John Brzorad, Keith Cooper, Robert Furness, Joel O'Connor, Fred Grassle, Larry Niles, and Ian Nisbet for their field companionship and valuable discussions. Several people helped review chapters, and I thank them as well: Caron Chess and Fred Grassle (Rutgers University), Michael Gochfeld (UMDNJ–Robert Wood Johnson Medical School), Rick B. Harley (Exxon), Carl Safina (National Audubon Society), and Dennis Suskowski (Hudson River Foundation). All of the authors also reviewed chapters, and I am grateful for their time and advice. I also thank Karen Reeds (science editor at Rutgers University Press) and Carole Brown (copyeditor) for their help and patience with the manuscript.

Contributors

John N. Brzorad
Ecology and Evolution
  Graduate Program
Rutgers University
Piscataway, NJ 08855

Joanna Burger
Biological Sciences
Rutgers University and
  Environmental and
  Occupational Health Sciences
  Institute
Piscataway, NJ 08855

Angela Cristini
Ramapo College
505 Ramapo Valley Rd.
Mahwah, NJ 07430

Keith R. Cooper
Toxicology Graduate Program
College of Pharmacy
Rutgers University
Piscataway, NJ 08855

Lynne Frink
TriState Bird Rescue
P.O. Box 289
Wilmington, DE 19801

Michael Gochfeld
Environmental and Community
  Medicine and Environmental
  and Occupational Health
  Sciences Institute
UMDNJ–Robert Wood Johnson
  Medical School
Piscataway, NJ 08855

Romona Haebler
United States Environmental
  Protection Agency
Environmental Research Lab
Narragansett, RI 02882

Paul M. Hauge
New Jersey Department of
  Environmental Protection
Office of Science Research
CN 409
Trenton, NJ 08625–0409

Gordon J. Johnson
New York State
Department of Law
120 Broadway, 26th floor
New York, NY 10271

Alan D. Maccarone
Environmental Studies
Friends University
Wichita, KS 67201

Katharine C. Parsons
Manomet Bird Observatory
Manomet, MA 02345

Carolyn Summers
Staff Analyst
New York City Department of
  Environmental Protection
New York, NY 10040

Robert K. Tucker
New Jersey Department of
  Environmental Protection
Office of Science Research
CN 409
Trenton, NJ 08625–0409

Before and After
an Oil Spill

1. Introduction

Joanna Burger

During the night of January 1–2, 1990, some 567,000 gallons of No. 2 fuel oil leaked into the Arthur Kill from a cracked underwater pipeline at the Exxon Bayway refinery and oil storage facility. The Arthur Kill is a short waterway that divides New Jersey and Staten Island (New York). ("kill" is Dutch for "creek" or "river.") It is part of the New York Harbor area, which includes the Upper and Lower Newark and Raritan bays. Although the Arthur Kill is heavily industrialized and has been exposed to numerous small oil spills, the massive leak of January 1–2 was an enormous exposure for such a small waterway.

This particular oil spill is unique among recent oil spills: It affected a well-defined, limited area within a large estuary; scientists had collected baseline data on biological systems in the Arthur Kill *before* the spill; comparable studies were made the year following the spill; the scientific assessments of environmental effects on the Arthur Kill and its food web served as the basis for a fast legal settlement with Exxon—within 2 years of the initial spill; and that settlement allows for open public and scientific scrutiny of the oil spill's consequences. We now have—as state officials, lawyers, scientists, wildlife rehabilitators, conservationists, industrialists, and citizens—a remarkable opportunity to learn from this spill.

In the aftermath of the spill meetings were held and coalitions formed among various parties including Exxon, New York and New Jersey state agencies, city governments, several conservation organizations, scientists, and the public. Immediate responses included cleanup operations and wildlife rescue and rehabilitation. These were followed by damage assessments and mediation. Decisions regarding courses of action in response to the spill were difficult because of the large number of key players, perhaps an inevitable feature of oil spills that occur in harbors, bays, estuaries, and other populated coastal areas. Damage assessments, however, were made easier because several scientists were already studying a variety of organisms in the Arthur Kill and baseline data were available on invertebrates, fish, and birds—key members of the food web of the Kill. Damage assessments were made by a number of scientists working for the government agencies of New York and New Jersey and several city and conservation agencies, as well as Exxon. These assessments served as the basis for the legal settlement with Exxon.

In this book we describe the events of the January 1–2, 1990, spill and subsequent cleanup and rescue operations. We examine the effects over the course of 2 years of the oil spill on plants and animals living in the Arthur Kill and review and evaluate the damage assessments provided

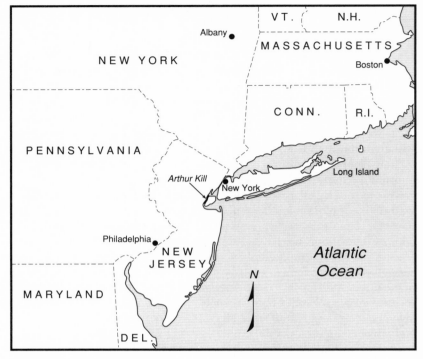

Figure 1.1. The Arthur Kill.

by scientists who worked for New York and New Jersey government agencies, several city and conservation agencies, and Exxon. Finally, and most important, we suggest ways to prevent, prepare for, and cope with future oil spills.

The Arthur Kill

The Arthur Kill, a tidal strait only 25 kilometers (15 mi.) long, separates Staten Island (New York) from northern New Jersey. The Arthur Kill is part of the New York–New Jersey harbor estuary (Fig. 1.1), a network of connecting tidal waterways located within eastern New York and northern New Jersey. This estuary receives freshwater drainage from an area encompassing 42,190 square kilometers (16,290 sq. mi.) (Rod, Ayres, and Small 1989) and is the largest source of fresh water for the New York bight, which extends from Montauk Point in New York south to Cape May, New Jersey.

This coastal area is perhaps the most used and abused coastal area in the world as a consequence of some twenty million people who reside by its shores (Swanson et al. 1990). New Jersey is the most densely populated state in the United States, and northern New Jersey is highly industrialized. New York City and its suburbs form a large metropolitan area. Thus, the Arthur Kill is surrounded by very dense populations, heavy industry, and major shipping lanes.

Because of its high population density and industrialization, New Jersey is often considered merely a corridor between New York City and Philadelphia, a land of turnpikes, parkways, industries, and airports. In recent years, however, New Jersey's natural habitats have attracted much attention, and the state and local governments have strongly asserted the importance of preserving natural areas in the midst of so much human activity (Pomper 1986). Two unique natural areas in particular, the Pine Barrens and the New Jersey shore, appeal to both residents and visitors as prime places for recreation and extended vacations. Their value is increased by their proximity to dense population centers. Guarded by the Pinelands Commission against encroachment and incompatible development, the New Jersey Pinelands are preserved for future generations. The New Jersey shore is partially protected from large-scale development, and its natural resources are guarded by a number of state agencies.

Likewise, in New York, Jamaica Bay Refuge (part of Gateway National Seashore) is close to metropolitan New York City and provides wild areas with extensive marshes and bays. The beaches on Long Island are highly protected. New York City also has experienced a regeneration of pride in its natural areas, culminating in countywide recognition of the beauties and uniqueness of the Big Apple.

With increasing interest in natural areas, people have turned to protect and appreciate local marshes, fields, woodlots, and forests. Not surprisingly, local protection is easy to engender for areas that are pristine and aesthetically pleasing, and whose birds and other animals are abundant. The presence of any human intrusion distracts us and overpowers our image of a natural ecosystem. We tend to view any habitat through our own eyes, seeing only the houses, landfills, industries, oil tanks, power lines, and abandoned cars, and fail to appreciate the value of the area to the other organisms that live there.

The January 1–2, 1990, oil spill in the Arthur Kill focused attention on the total ecosystem surrounding the Kill, not just the industrialization and other human activities. With full media fury, all aspects of life in the Arthur Kill were examined by government officials, by industry management, by scientists, by people who live and work along the Kill, and by the public.

The Arthur Kill Ecosystem

Physical Features

The Arthur Kill connects the Kill van Kull and Newark Bay (and the Hackensack River) to the north, and Raritan Bay to the south (see Fig. 1.1). Major tributaries of the Arthur Kill are the Rahway and Elizabeth rivers, and Fresh Kills Creek. Because this area is highly industrialized and densely populated, little remains of the original waterway. Except for the Hackensack River, the Kill complex has few natural banks left. Much is bulkheaded or otherwise developed for piers, docks, and other industrial uses.

The Arthur Kill is located in the Piedmont Lowlands in the north (with Triassic rocks) and the Coastal Plain in the south (with Cretaceous rocks) (Johnson 1950; Wolfe 1976). Its sediments are composed largely of silt (60%), with some clay (25%), and sand (18%) (Ichthylogical Associates, 1976). Mud flats still line some parts of the Arthur Kill, and the main channel bottom averages 30 feet in depth as a result of dredging (Entrix and International Technology 1990).

The Rahway River drains approximately 2,135 square kilometers (83 sq. mi.) yet discharges fresh water at an average rate of 2,435 liters (86 cu. ft.) per second; the Elizabeth River drains 54 square kilometers (21 sq. mi.) and discharges fresh water at an average annual rate of 680 liters (24 cu. ft.) per second; and the Fresh Kills Creek discharges fresh water at an average rate of only 340 liters (12 cu. ft.) per second (Hires 1988). Fresh water, however, also enters the Arthur Kill from municipal sewage treatment plants, sewer outfalls (including rainfall), and industrial discharges (Hires 1988). These three sources may account for about 5,664 liters (200 cu. ft.) per second, or 62 percent of the fresh water entering the Arthur Kill.

The continual intermixing of tidal waters with fresh water from the rivers and from anthropogenic sources results in variable salinity, ranging from 17 to 27 parts per thousand at the southern end of the Kill to almost 0 in the tributaries. Organisms living within the Kill have adapted to withstand rapid shifts in salinity.

The Arthur Kill is unusual in that it receives tidal waters from each end. Tides surge into it from both the north, through the Kill van Kull and Newark Bay, and from the south, through Raritan Bay. The southern ebbing tide is approximately 36 minutes longer than the flooding tides, indicating a net transport from the Kill to Raritan Bay (Entrix and International Technology 1990). Others, however, argue that the net transport is to the north. Tidal range is 1.3 to 1.9 meters (4.2 to 6.4 ft.) with a mean of 1.6 meters (5.3 ft.) (Oey, Mellor, and Hires 1985). The tidal surges from both directions cancel out to some extent, and the water often flows back and forth within the Kill for some time before leaving. On average, water stays in the Kill for 2 weeks or more (Oey, Mellor, and Hires 1985). This compounds the effects of contaminants, because they are not rapidly washed out to sea by the tides.

Primary productivity and the food web ultimately depend on water quality. Dissolved oxygen, in particular, is critical to aquatic life and to the natural decomposition processes. Without dissolved oxygen the food chain collapses at all but anaerobic levels. Historically, much of the New York–New Jersey harbor estuary was degraded because of low dissolved-oxygen levels caused by the large-scale dumping of raw sewage into these waters (Loop 1964; Suszkowski 1990). From the 1960s through the 1980s, water quality improved with the general cessation of sewage disposal, and dissolved oxygen increased, bacteria decreased, lead decreased, and ammonia decreased (NYCDEP 1987).

The slow tidal flushing in the Arthur Kill contributes markedly to water-quality problems. Because the water remains in the Kill for long periods of time, pollutants also remain for long periods. Spatial and temporal variations in water quality in the Kill are great (Brosnan, Stokes, and Forndran 1987). Although nutrient levels are high, algae blooms have not been reported, perhaps because of turbidity or because of the high levels of toxics and ammonia. Levels of heavy metals such as mercury, zinc, cadmium, and lead are generally high in the harbor estuary and are particularly high in the Arthur Kill itself (Meyerson 1988; EA Engineering, Science and Technology 1989; Environmental Consulting Services 1989; Squibb, O'Connor, and Kneip 1990). Grieg and McGrath (1977) mapped the extent of heavy-metal contamination in Raritan Bay and found the highest concentrations of several metals (indeed some of the highest concentrations in the world) at the point where the Arthur Kill enters Raritan Bay.

Heavy metals have a variety of effects on marine organisms (Burger, Parsons, and Gochfeld in press). For example, in birds, lead causes decreased egg production and decreased sperm count (Stone and Soars 1974; Kendall, Veif, and Scanlon 1981); mercury causes decreased egg production and thin shells (Spann et al. 1972); nickel causes developmental defects (Gilani and Marano 1980); and cadmium causes decreased testes weight (White and Finley 1978).

Other pollutants such as DDT (dichlorodiphenyl trichloroethane) and its breakdown products cause thin eggshells in birds (Peakall and Lincer 1972). In other organisms these same pollutants also cause developmental, reproductive, and physiological damage (see Newman and McIntosh 1991; Weis and Weis 1991). Polychlorinated biphenyl compounds (PCBs) are sufficiently high in the Arthur Kill to warrant advisories concerning consumption of fish and shellfish (Belton, Ruppel, and Lockwood 1982). It is on this tapestry of imperfect water quality that petroleum hydrocarbons spill.

Biological Features

Any ecosystem can be viewed from either a structural or a functional viewpoint. A structural analysis examines the trophic-level relationships, including the plants, zooplankton, benthos communities, fish, and birds. A functional analysis examines the process of community interactions, such as competition, predation, herbivory, commensalism, and reproductive processes. A healthy ecosystem has a diverse structure and a complex interplay of functions.

The physical features of the Arthur Kill include a waterway of varying depths, tidal mud flats, tidal creeks, and salt marshes that grade into upland areas supporting herbs, shrubs, and trees. The banks and associated mud flats and marshes of the Arthur Kill are highly modified by human interests. The shores include bulkheads, riprap, sand, and gravel, as well as marshes and mud flats. Along the New Jersey side, the linear shoreline is only 6 percent mud flats and 32 percent marshes, compared with 22 percent mud flats and 51 percent marshes along the New York side (USACOE 1980). Thus, overall only 55 percent of the shoreline (including the islands) is natural mud flats and marshes. Considering that the Arthur Kill flows through the most highly industrialized harbor area in the United States, it is remarkable that any natural shoreline remains.

The marshes, uplands, mud flats, shallows, and tidal waters are the major habitats that make up the Arthur Kill. The salt marshes are developed mainly on marine sediments and marsh peat. The low marsh, closest to the water and exposed to the most frequent tidal inundation, is primarily composed of cordgrass (*Spartina alterniflora*), while the higher places are dominated by salt hay (*S. patens*). The distribution of *Spartina* is influenced by substrate, immersion time, and salinity (Teal and Teal 1969; Teal 1986). Algae contribute significantly to the primary productivity of the marsh and constitute an important food source for many salt marsh and estuarine organisms.

The most extensive salt marshes of the Arthur Kill occur on Staten Island. These include the marshes surrounding Old Place Creek (across from the Exxon spill of January 1–2, 1990), Sawmill Creek, Neck Creek, Little Fresh Kills, Great Fresh Kills, Main Creek, and Richmond Creek. Some salt marshes are found on Isle of Meadows and Pralls Island.

Salt marshes serve a number of important functions within the Arthur

Kill ecosystem; providing primary productivity, nutrient cycling, and habitat (Teal and Teal 1969). Plant tissue and algae produced on the marsh are broken down into detritus, which is then transported to other areas. This detritus provides dissolved and particulate organic carbon, dissolved organic nitrogen, and dissolved phosphorus, which are essential sources of energy and nutrients for the entire ecosystem. Furthermore, the high organic-matter inputs from sewage affect nutrient flows. Salt marshes also serve as nursery areas for many species of fish and shellfish.

Many species live in the intertidal zone. These include fiddler crabs (*Uca* spp.), ribbed mussels (*Geukensia demissus*), and marsh snails (*Melampus bidentatus*) (Franz 1973, 1982). A wide diversity of other invertebrates live in association with these species (Teal and Teal 1969) and are essential components of the food web. Fish and birds depend upon the salt marsh for foraging (Maccarone and Parsons 1988; Maccarone 1989). Indeed, the herons, egrets, ibises, and gulls that breed on the islands in the Kill rely extensively on the salt marsh and mud flats, as well as on freshwater ponds and tributaries, for their prey throughout the breeding cycle.

On still higher places the salt marsh gives way to uplands, with grasses, herbs, shrubs, and eventually trees. Upland areas on Pralls Island, Isle of Meadows, and Shooters Island provide nesting places for a variety of avian species including herons, egrets, ibises, and gulls (Parsons 1989; Burger, Parsons, and Gochfeld in press). These higher places, created by the deposition of dredge spoil, are not generally exposed to tidal inundation except during hurricanes and other severe storms. Even then the degree of inundation depends on the elevation of the land and the severity of the winds. Because it was easily developed, upland habitat was traditionally the first type of terrain used by people. This left little natural upland area, increasing the importance of the limited upland on Pralls and Shooters islands, and on Isle of Meadows.

The Arthur Kill is used extensively by a community of bird species that breed in the upland areas. The three islands in the Arthur Kill currently provide nesting habitat for the largest heronry in New York State (Parsons 1989), and in 1989 accounted for more than 30 percent of New York's Ciconiiformes (herons, egrets, and ibises) (Downer and Liebelt 1989, 1990). This is all the more remarkable because these species began nesting in the Kill only in the mid 1970s (Buckley and Buckley 1980).

The uplands however, also provide nesting areas for colonies of herring and great black-backed gulls (*Larus argentatus* and *L. marinus*), as well as for solitary-nesting ducks and geese. At least one marsh hawk pair (*Circus cyaneus*) nests in the upland–salt marsh fringe, and many hawks overwinter in the Kills area (Burger, Parsons, and Gochfeld in press).

Below the salt marsh is the intertidal mud-flat community. Invertebrates in this community include a variety of animals including sponges, jellyfish, worms, shellfish, and snails. The distribution of benthic, or sediment-dwelling, organisms is partly determined by the character of the substrate. The intertidal substrate of the Arthur Kill varies from solid artificial surfaces (concrete, wooden piers, bulkheads), to rocks and pebbles, to silt and mud. Since intertidal invertebrates are aquatic, exposure to air at low tides is the most important factor affecting their survival. These intertidal species exhibit zonation along a tidal gradient, which determines their relative exposure to pollutants such as oil. The more tidal inundation they are exposed to, the greater their contamination with oil. The majority of invertebrates in the Arthur Kill are jellyfish, hydroids, polychaete worms, snails, clams, crabs, shrimp, and amphipods (Entrix and International Technology 1990). Generally, benthic fauna diversity is low in the Arthur Kill (Entrix and International Technology 1990).

Natural environmental stress, toxics, and human disturbance all influence the presence, abundance, and distribution of benthic invertebrates. High population levels of certain polychaete worms are often associated with pollution or disturbance. Nematode worms, polychaete worms, and tubificid worms, all of which are common in the Arthur Kill (Mayer 1982), are associated with near-anoxic conditions (Tietjen 1982; Young and Young 1982). Some amphipods that are missing from the New York bight area are those that are sensitive to pollutants (Lee et al. 1977; Young and Young 1982). Petroleum hydrocarbons cause mortality in some invertebrates (Anderson 1982; Mayer 1982). Thus, a detailed census of benthic invertebrate fauna can reveal the absence of species sensitive to pollutants, the presence of species that indicate pollutants or environmental stress, or the high species diversity indicative of a healthy ecosystem.

The phytoplankton and zooplankton communities span habitats from the mud flats to deep waters of any coastal ecosystem. The phytoplankton community is the primary producer of the ecosystem, using chlorophyll-a in the photosynthetic process. Therefore, chlorophyll-a is often a measure of primary productivity. Productivity, in terms of chlorophyll-a, decreases from Raritan Bay north to the Kill van Kull (NYCDEP 1987; NJDEP 1988; EA Engineering, Science and Technology 1989). In a system that receives a high nutrient load from sewage, as the Kill does, this aspect of productivity needs to be included in the ecosystem models of the region.

Light, nutrients, and the presence of pollutants affect primary productivity. Low light is a limiting factor for phytoplankton in the Arthur Kill (whose waters are turbid) (Malone 1982). Since nutrient (nitrogen and phosphorus) input into the Arthur Kill is high because of some dumping of sewage, nutrients far exceed the demands of the phytoplankton (Mayer 1982). Diatoms are the largest phytoplankton group in the Arthur Kill (NYCDEP 1987) and are critical as food for clam populations (McLaughlin et al. 1982). Primary productivity in the Arthur Kill may also be limited by pollutants.

Zooplankton include a variety of adult and larval invertebrates, larval fish, and fish eggs. Zooplankton graze on phytoplankton, bacteria, or smaller zooplankton and are themselves a food source for secondary consumers such as larger invertebrates and fish. Resources are partitioned among zooplankton, and a daily vertical migration occurs among some species: Some zooplankton rise toward the surface at night and descend to deeper water at dawn. Since many feed on phytoplankton, their distribution is often determined by the phytoplankton (Wetzel 1983).

Zooplankton populations usually peak after phytoplankton populations. These population peaks are often accompanied by spawning, hatching, or breeding activities among other invertebrates, fish, and marine birds that feed on them (Lerman 1986).

In the Arthur Kill the most abundant microzooplankton are crustaceans (EA Engineering, Science and Technology 1989), while the larvae of mud crabs and grass shrimp (*Palaemonetes*) are the dominant macrozooplankton (Raytheon 1972). The eggs and larvae of thirty-six species of fish have been collected in the Arthur Kill, but abundance is lower than nearby areas (Ichthyological Associates 1976; EA Engineering, Science and Technology 1989). Thus, the bulk of the fish larvae and fish eggs in the Arthur Kill are believed to originate in Raritan Bay or Lower New York Bay (Entrix and International Technology 1990).

The fish community in the Arthur Kill is composed of permanent, year-round residents, seasonal migrants, and accidental or irregular visitors. Over sixty species of fish have been collected in the Arthur Kill over the past 20 years (Howells and Brundage 1977). Because of increasing dissolved-oxygen levels over the last 20 years, fish diversity and abundance has increased, perhaps creating a more complex trophic web in the Arthur Kill.

Mummichog (*Fundulus heteroclitus*) and grubby sculpin (*Myxocephalus aeneus*) are present all year. Mummichogs form the basis for the next trophic level, that of larger fish and the avian community (Berg and Levinton 1985; Brzorad and Burger 1989). Several other species such as anchovy (*Anchoa mitchilli*), Atlantic silverside (*Menidia menidia*), and alewife (*Alosa pseudoharengus*) are migratory. Some predatory fishes such as bluefish (*Pomatomus saltatrix*), weakfish (*Cynoscion regalis*), and hakes (*Urophycis* spp.) are also found in the Arthur Kill during migration or when young move from the spawning grounds to the sea.

Kennish (1990) recognizes two types of food webs in coastal systems: detrital and grazing. The detrital food web (bacteria, fungi, microalgae) processes over 90 percent of the primary productivity (from *Spartina* and macroalgae) (Knox 1986). The detrital particles are then consumed by demersal fish, mollusks, and polychaete worms (Kennish 1990). In the grazing food web, the primary productivity biomass is consumed directly by zooplankton, planktivorous fish, omnivores, microfauna, meiofauna, macroinvertebrates, surface feeders, and filter feeders (Knox 1986). In both food webs the herbivores are consumed by carnivores, which include primarily predatory fish and predatory birds.

This brief overview of the structure of the Arthur Kill ecosystem provides a basis for the chapters that follow, with their discussion of the effects of oil on specific organisms and communities. The diversity and structure of the Arthur Kill ecosystem affects the diversity and complexity of the trophic relationships, or food web, within the Kill. Obviously, a disruption to the primary producers affects the zooplankton and primary consumers, and, ultimately, the entire food web (Kennish 1990).

The Exxon Oil Spill of 1990

On January 1–2, 1990, 567,000 gallons of No. 2 fuel oil appeared on the surface of the Arthur Kill. The crack in the 12-inch underwater pipeline that connects the Exxon Bayway refinery at Linden, New Jersey, to the Bayonne plant in Bayonne, New Jersey, was confirmed by divers. The pipeline leak was just south of the Goethals Bridge (Fig. 1.2 and 1.3). Tides and winds quickly moved the oil to the Staten Island side of the Kill and to the three islands in the Kills area (Pralls, Shooters, and Isle of Meadows). Eventually it spread to both shores. The extent of the oil spill is shown in Figure 1.3.

The spill was considered major by the Coast Guard because it involved more than 10,000 gallons. Cleanup efforts were conducted by Exxon and were monitored by the Coast Guard and the Atlantic Strike Team. The cleanup effort involved 60,000 feet of boom, 680 people, 70 boats, 40 vacuum trucks, and 10 skimmers (Fig. 1.4). The cleanup resulted in the recovery of 141,000 gallons (25%) of the oil. Exxon estimated that 50 percent of the oil product evaporated. This leaves at least 142,500 gallons unaccounted for, but a significant part must have permeated the sediments, mud flats, and marshes of the Kill or been washed out by tides to the New York–New Jersey estuary and the Atlantic Ocean. The cleanup operation ended on March 15, 1990. A more de-

Figure 1.2. Site of the pipeline leak.

tailed description of the cleanup operations and decisions about these operations can be found in Chapter 2.

The Exxon oil spill is remarkable because of its magnitude and the resultant industry, governmental, and public response. Yet oil pollution in the Arthur Kill is a continuous problem because of the concentration of oil refineries, loading docks for oil tankers, and storage tanks along its banks. Waste petroleum also occurs in municipal sewage systems because it cannot be prevented from entering the systems from roads, parking lots, and households (Tanacredi and Maltezou 1980; Tanacredi 1990).

Some seventeen hundred oil tankers enter the New York–New Jersey harbor estuary each year (Natural Resources Defense Council 1990), and 18 billion gallons of oil arrive in the harbor annually. Most of the storage and transfer facilities are concentrated in the Kill region (Fig. 1.5), increasing the density of the traffic and the potential for problems. The average tanker carries 13 million gallons of oil, although some hold as much as 25 million gallons. Barges carry up to 8 million gallons. An average of 100 oil transfers a day (totaling 50 million gallons) take place (Natural Resources Defense Council 1990). About 3 billion gallons of oil are stored in the Arthur Kill–Kill van Kull, compared with just over 500 million gallons for the rest of the harbor.

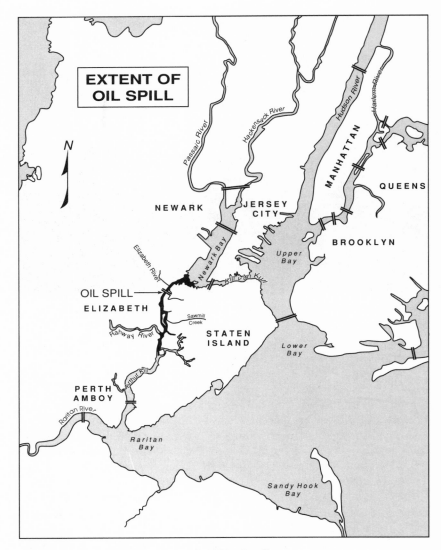

Figure 1.3. Extent of the January 1–2, 1990, oil spill.

With the number of transport vessels, the quantities of oil being transported and transferred, and the number of underground pipelines, the potential for leaks, collisions, groundings, and fires is great, as are the chances for oil spills. On average, oil spills occur almost daily in the New York–New Jersey harbor (Natural Resources Defense Council

Figure 1.4. One of the cleanup areas, showing booms removed from salt marsh creeks.

1990), but they are usually small, contained, and go unnoticed by the public. The number of oil spills increased from 1987 to 1990 but decreased in 1991 (Natural Resources Defense Council 1990; Burger, Parsons, and Gochfeld in press).

Year	Number of incidents
1982	118
1983	73
1984	65
1985	65
1986	78
1987	77
1988	182
1989	275
1990	261
1991	177

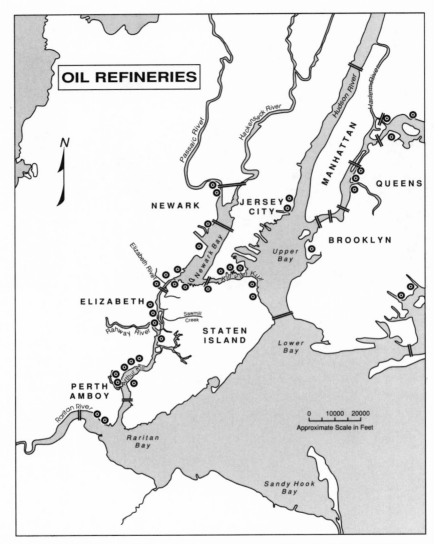

Figure 1.5. Concentration of oil storage and transfer facilities in the New York–New Jersey harbor estuary.

Unfortunately, the extent and quantities of petroleum hydrocarbon pollution in the Arthur Kill and the rest of the New York–New Jersey harbor estuary are not regularly monitored, although the Coast Guard monitors large spills. Such monitoring would be useful both to assess for regulatory purposes levels of oil spilled as well as to correlate with the biological resources in the Arthur Kill. Although the U.S. Department of the Interior (1967) reports that oil is frequently observed on the surface waters, no analyses are usually conducted of routine small spills. In one study Mytelka et al. (1981) reported that petroleum hydrocarbon levels were ten times as high in the Hudson–Raritan estuary as in the open Atlantic Ocean, and levels in the Arthur Kill were among the highest in the world (organics of 138 to 255 μg/l and hydrocarbons of 25 to 73 μg/l).

As the chapters that follow indicate, the organisms in any tidal estuary have, over many thousands of years, evolved mechanisms to withstand changing environmental conditions such as devastating floods, temperature and tidal shifts, salinity fluctuations, disease, and interspecific interactions (such as competition and predation). To some extent these mechanisms have preadapted organisms to withstand the insults of environmental pollution, including oil. Natural populations and ecosystems require time to recover, however. If environmental insults occur repeatedly, organisms can no longer recover between the events, and populations decline, species become extirpated, and the complex interactions of a healthy ecosystem break down.

In this book we examine the responses of a variety of organisms and ecosystems to the Exxon spill in the Arthur Kill and to oil spills in general. We also explore the responses of the public, scientists, and government agencies to the spill. Our overall objective is to understand the effect of the oil spill on the Arthur Kill ecosystem, including humans, and to describe how the social and political system responded to the spill. This volume should provide a basis for all coastal communities in preparing for, and understanding, oil spills.

Conclusions

The Arthur Kill is located in one of the most densely populated parts of the United States, yet it harbors a tidal ecosystem with salt marshes and a complex food web culminating in three breeding colonies of gulls, herons, egrets, and ibises. The Exxon oil spill of January 1990 dumped more oil at one time into the Arthur Kill than has any other single event. Subsequently, cleanup and damage assessment were initially hampered by the number of state agencies (from both New York and New Jersey),

federal agencies, and private organizations involved, although cooperation normally prevailed.

The events that followed the Arthur Kill oil spill provide valuable
information about the ways both natural and human communities respond to industrial accidents. From these responses we can learn how to
protect all ecosystems and also gain valuable knowledge on how to best
organize ourselves before and after such times of crisis. Future decisions
about cleanup, damage assessment, biological monitoring, and subsequent rehabilitation will profit from both an understanding of the human
and biological resources of an area and an evaluation of outcomes from
past decisions, such as those that followed the Exxon oil spill in the
Arthur Kill.

Acknowledgments

I thank John Brzorad, Caron Chess, Michael Gochfeld, R.B. Harley,
Karen Reeds, and Dennis Suszkowski for valuable comments on the
manuscript. Robert Tucker and Paul Hauge provided access to many of
the documents from the Arthur Kill oil spill damage assessment.

Literature Cited

Anderson, J. W. 1982. The transport of petroleum hydrocarbons from sediments
    to benthos and the potential effects. In G. F. Mayer, ed., *Ecological Stress
    and the New York Bight: Science and Management*, pp. 165–180. Columbia,
    S.C.: Estuarine Research Federation.
Belton, T. J., B. E. Ruppel, and K. Lockwood. 1982. *PCBs (Aroclor 1244) in
    Fish Tissues Throughout the State of New Jersey: A Comprehensive Study.*
    Trenton: New Jersey Department of Environmental Protection.
Berg, D. L., and J. S. Levinton. 1985. The biology of the Hudson–Raritan
    estuary, with emphasis on fishes. NOAA technical memorandum NOS OMA
    16.
Brosnan, T. M., T. L. Stokes, and A. B. Forndran. 1987. Water quality monitor-
    ing and trends in NY Harbor. Proceedings of the Oceans 87 Conference, vol.
    5, Halifax, N.S., September 30, 1987.
Brzorad, J. N., and J. Burger. 1989. The influence of tide on the abundance of
    egrets and their prey in a New Jersey estuary. *Colonial Waterbird Society
    Newsletter* 13:2.
Buckley, P. A., and F. G. Buckley. 1980. Population and colony site trends of
    Long Island waterbirds for five years in the mid 1970's. *Trans. Linn. Soc. N.Y.*
    9:23–56.
Burger, J., K. Parsons, and M. Gochfeld. In press. Avian populations and envi-

ronmental degradation in an urban river: The kills of New York and New Jersey. In J. A. Jackson, Jr., ed., *Avian Conservation*. Madison: University of Wisconsin Press.

Downer, R. H., and C. E. Liebelt. 1989. *The 1989 Long Island Colonial Waterbird and Piping Plover Survey*. Stony Brook, N.Y.: New York Department of Environmental Conservation.

————. 1990. *The 1989 Long Island Colonial Waterbird and Piping Plover Survey*. New York Department of Environmental Conservation.

EA Engineering, Science and Technology. 1989. Linden generating station. Supplemental 316(a), demonstration and supplemental 316(b) report. Prepared for Public Service Electric and Gas Co., Newark, N.J.

Entrix and International Technology. 1990. The environment and communities of the Arthur Kill. Report to Exxon Co., October 1990.

Environmental Consulting Services, Inc. 1989. A biological characterization of Morses Creek: A summary. Prepared for Exxon Co. U.S.A.

Franz, D. R. 1973. The ecology and reproduction of a marine bivalve, *Myrella planulata* (Erycinacea). *Biol. Bull.* 144:93–106.

————. 1982. An historical perspective on mollusks in lower New York Harbor, with emphasis on oysters. In Mayer, *Ecological Stress*, pp. 181–198.

Gilani, S. H., and M. Marano. 1980. Congenital abnormalities in nickel poisoning in chick embryos. *Arch. Environ. Contam. Toxicol.* 9:17–22.

Grieg, R. A., and R. A. McGrath. 1977. Trace metals in sediments of Raritan Bay. *Marine Poll. Bull.* 8:188–191.

Gunster, D. G., C. G. Gillis, N. L. Bonnevie, T. B. Abel, and R. J. Wenning. 1993. Petroleum and hazardous chemical spills in Newark Bay, New Jersey, USA, from 1982 to 1991. *Environ. Pol.* 82:245–253.

Hires, R. 1988. Circulation and transport in the Hudson–Raritan estuary. In *The Hudson–Raritan: State of the Estuary*. Appendix, vol. 1, pt. 2 of *Water Quality of the New Jersey Coastal Waters*. Fort Hancock, N.J.: New Jersey Marine Sciences Consortium.

Howells, R. G., and H. M. Brundage III. 1977. Fishes of the Arthur Kill. *Proceedings of the Staten Island Institute of Arts and Sciences* 29:3–6.

Ichthyological Associates, Inc. 1976. Ecological studies of the Arthur Kill and Morses Creek in the vicinity of the Exxon Bayway refinery, Linden, NJ. May 1975–1976, pts. 1 and 2. Prepared for Exxon Co., U.S.A.

Johnson, M. E. 1950. Geological map of New Jersey. New Jersey Department of Geology and Topography.

Kendall, R. J., H. P. Veif, and P. F. Scanlon. 1981. Histological effects and lead concentrations in tissues of adult male ringed turtle doves that ingested lead shot. *J. Toxicol. Environ. Health* 8:649–658.

Kennish, M. J. 1990. *Ecology of Estuaries*. Vol. 2, *Biological Aspects*. Boca Raton, Fl.: CRC Press.

Knox, G. A. 1986. *Estuarine Ecosystems: A Systems Approach*, vol. 1. Boca Raton, Fl.: CRC Press.

Lee, R., A. C. Longwell, T. C. Malone, L. S. Murphy, D. R. Nimmo, H. B.

O'Connors, L. S. Peters, and K. D. Wyman. 1977. Effects of pollutants on plankton and neuston. In Mayer, *Ecological Stress*, pp. 39–52.

Lerman, M. 1986. *Marine Biology*. Menlo Park, Calif.: Benjamin/Cummings.

Loop, A. S. 1964. *History and development of sewage treatment in New York City*. New York City Department of Health.

Maccarone, A. D. 1989. Factors afffecting the use of foraging habitats by breeding herons, egrets, and ibises. *Bull. N.J. Acad. Sc.* 34:36.

Maccarone, A. D., and K. C. Parsons. 1988. Differences in flight patterns among nesting ibises and egrets. *Colonial Waterbirds* 11:67–71.

McLaughlin, J.J.A., G. S. Kleppel, M. P. Brown, R. J. Ingram, and W. B. Samuelts. 1982. The importance of nutrients to phytoplankton production in New York Harbor. In Mayer, *Ecological Stress*, pp. 469–480.

Malone, T. C. 1982. Factors influencing the fate of sewage-derived nutrients in the lower Hudson estuary and New York bight. In Mayer, *Ecological Stress*, pp. 389–400.

Mayer, G. F. 1982. *Ecological Stress and the New York Bight: Science and Management*. Columbia, S.C.: Estuarine Research Federation.

Meyerson, A. L. 1988. Contaminants in the sediments of the Hudson–Raritan estuarine system. In *The Hudson–Raritan*. Appendix, vol. 1, pt. 2 of *Water Quality of New Jersey Coastal Waters*.

Mytelka, A. I., M. Wendell, P. L. Sattler, and H. Golub. 1981. *Water Quality of the Hudson–Raritan Estuary*. Boulder, Colo.: National Oceanic and Atmospheric Administration, Office of Marine Pollution Assessment. (Prepared by Interstate Sanitation Commission, New York.)

Natural Resources Defense Council. 1990. *No Safe Harbor: Tanker Safety in America's Ports*. New York: Natural Resources Defense Council.

New Jersey Department of Environmental Protection (NJDEP). 1988 (May). New Jersey 1988 state water quality inventory report. Trenton, N.J.

Newman, M. C., and A. W. McIntosh. 1991. *Metal Ecotoxicology: Concepts and Applications*. Chelsea, Mich.: Lewis.

New York City Department of Environmental Protection, Bureau of Wastewater Treatment (NYCDEP). 1987. *New York Harbor Water Quality Survey, 1987*. New York: New York City Department of Environmental Protection, Bureau of Wastewater Treatment, Water Quality Section.

Oey, I. Y., G. L. Mellor, and R. L. Hires. 1985. Tidal modeling of the Hudson–Raritan estuary. *Estuarine, Coastal and Shelf Science* 20:511–527.

Parsons, K. C. 1989. *The Harbor Herons Project Interpretive Report*. Manomet, Mass.: Manomet Bird Observatory.

Peakall, D. B., and J. L. Lincer. 1972. Methylmercury: Its effect on eggshell thickness. *Bull. Environ. Contam. Toxicol.* 8:89–90.

Pomper, G. M., ed. 1986. *The Political State of New Jersey*. New Brunswick, N.J.: Rutgers University Press.

Raytheon Co. 1972. *An Ecological Study of the Arthur Kill*. Wayne, N.J.: Raytheon Co., Environmental System Center, Environmental Research Laboratory,

Rod, S. R., R. U. Ayres, and M. Small. 1989. Reconstruction of historical loadings of heavy metals and chlorinated hydrocarbon pesticides in the Hudson–Raritan basin, 1880–1980. Final report to the Hudson River Foundation, New York.

Spann, J. W., R. G. Heath, J. F. Kreitzer, and L. N. Locke. 1972. Ethyl mercury p-toluene sulfonanilide: Lethal and reproductive effects on pheasants. *Science* 175:328–331.

Squibb, K. S., J. M. O'Connor, and T. J. Kneip. 1990 (October). New York/New Jersey harbor estuary toxics categorization. Draft report prepared for U.S. Environmental Protection Agency Region II.

Stone, C., and J. H. Soars. 1974. Abstract: Studies on the metabolism of lead in Japanese Quails. *Poultry Science* 53:1982.

Suszkowski, D. J. 1990. Conditions in the New York–New Jersey harbor estuary. In M. T. Sutherland, ed., *Cleaning Up Our Coastal Waters: An Unfinished Agenda*, New York: U.S. Environmental Protection Agency.

Swanson, R. L., T. M. Bell, J. Kahn, and J. Olha. 1990. Use impairments and ecosystem impacts of the New York bight. In Sutherland, Cleaning Up Our Coastal Waters, pp. 133–170.

Tanacredi, J. T. 1990. Napthalenes associated with treated wastewater effluent in an urban national wildlife refuge. *Environ. Cont. and Toxicol.* 44:246–253.

Tanacredi, J. T., and F. Maltezou. 1980. Waste oil recycling: Are there any incentives out there? Proceedings of the 1980 Year of the Coast Conference. Dowling College, N.Y.

Teal, J., and M. Teal. 1969. *Life and Death of the Salt Marsh.* Boston, Mass.: Little, Brown.

Teal, J. M. 1986. The ecology of regularly flooded salt marshes of New England: A community profile. *U.S. Fish Wild. Serv. Biol. Rep.* 85.

Tietjen, J. H. 1982. Potential roles of nematodes in polluted ecosystems and the impact of pollution on meiofauna. In Mayer, *Ecological Stress*, pp. 225–234.

U.S. Army Corps of Engineers (USACOE). 1980. EIS for Kill van Kull and Newark Bay Channels, New York and New Jersey. U.S. Army Corps of Engineers, New York District.

U.S. Department of the Interior (USDOI). 1967. Summary report for the conference of pollution of Raritan Bay and adjacent interstate waters (third session). Federal Pollution Control Administration, Northeast Region—Raritan Bay Project.

Weis, P., and J. W. Weis. 1991. The developmental toxicity of metals and metalloids in fish. In M. C. Newman and A. W. McIntosh, eds., *Metal Ecotoxicity: Concepts and Applications*, pp. 145–170.

Wetzel, R. G. 1983. *Limnology*, 2d ed. New York: CBS College Publishing.

White, D. H., and M. T. Finley. 1978. Uptake and retention of dietary cadmium in mallard ducks. *Environ. Res.* 17:53–59.

Wolfe, P. E. 1977. *The Geology and Landscapes of New Jersey.* New York: Crane, Russak.

Young, M. W., and D. K. Young. 1982. Marine macrobenthos as indicators of environmental stress. In Mayer, *Ecological Stress*, pp. 527–539.

Responses and Cleanup

2. Governmental Cooperation

Paul M. Hauge and Robert K. Tucker

When approximately 567,000 gallons of no. 2 fuel oil poured out of an underwater rupture in the Exxon Company's Inter-refinery Pipe Line into the Arthur Kill in the vicinity of Morses Creek during the night and morning of January 1–2, 1990, it triggered responses by agencies of the federal government, two states, and at least two municipalities. While any oil spill will likely involve more than one government agency, the location of the Exxon spill—in interstate waters, near natural resources held in trust by two states and the federal government, close to wildlife populations of regional significance, and on the doorstep of the nation's largest city—meant the involvement of an unusually large number of agencies and individual offices.

Since the spill area was bordered by two states, the need for interagency coordination and cooperation in immediate spill response activities (and hence for preexisting agreements and working relationships) was intuitively obvious. In the case of the 1990 Exxon spill in the Arthur Kill the response effort was further complicated by the fact that authorities were not immediately notified of the spill and its location. This delayed the response and contributed to the eventual spreading of the oil to much of the Arthur Kill and its tributaries, as well as to adjacent estuarine waters in the Kill van Kull and Newark Bay. Coordination was important during cleanup operations, as authorities from the numerous agencies involved determined appropriate measures and decided when further efforts were either unwise or unnecessary. Even more complicated, and certainly lengthier, than the response and cleanup efforts was the process by which the various government agencies sought to assess the effects of the spill on the much-maligned resources of the Arthur Kill so as to meet the needs (and timetables) of those charged with carrying out their respective legal responsibilities.

Soon after the spill, authorities from the three levels of government determined that a coordinated legal and technical response to the spill was necessary. This chapter describes the efforts of agency personnel from the federal government, New Jersey, New York State, New York City, and the City of Elizabeth to assess the effects of the spill and to contribute to the resolution of the various legal actions brought against Exxon.

Legal Context for Oil Spills

Within weeks after the spill, several of the governments had filed lawsuits against Exxon. Among the legal claims against the company were those for monetary damages for injuries to natural resources affected by the spill. The laws and regulations concerning such damages provided the context for all discussions concerning scientific assessments of the spill's effects.

The concept of monetary damages for injuries that a spill or other discharge causes to natural resources has its roots in the public trust doctrine, that is, the principle that natural resources such as wildlife, shellfish beds, and waterways are held in trust by the government on behalf of the public. Various common-law theories can support actions for natural-resource damages, as can statutes in a number of states, including New Jersey (Spill Compensation and Control Act and Water Pollution Control Act). In the area of spills and discharges, the concept is also codified in two federal statutes, the Comprehensive Environmental Response, Compensation, and Liability Act (CERCLA, also known as the Superfund Law), and the Clean Water Act. The Clean Water Act covers discharges to surface and groundwater. CERCLA covers discharges of hazardous substances but not petroleum products. Both federal statutes name representatives of state governments, the federal government, and native-American tribes as trustees for natural resources. This gives these officials the right to sue spillers to recover monetary damages for the destruction of or injury to natural resources caused by the spill.

CERCLA also requires the federal government to develop regulations governing how natural-resource trustees can assess natural-resource damages. These regulations are to apply to spills covered by both CERCLA and the Clean Water Act. Trustees do not necessarily have to follow the procedures set out in such regulations, but if they choose not to they lose the legal presumption that the assessment is correct. CERCLA calls for regulations for two types of assessments, named for the subparagraphs that refer to them: Type A, for simplified assessments covering most minor spills, and Type B, for more detailed assessments required for large or especially harmful spills.

Almost from their inception (1976), the CERCLA regulations were subject to controversy. They were drafted only after several lawsuits were filed and were eventually adopted in 1986 (Type B) and 1987 (Type A). The regulations were then challenged in court by environmental groups and ten states. In two separate decisions reached in 1989, *Ohio v. United States Department of the Interior* (Type B) and *Colorado v.*

United States Department of the Interior (Type A), the Circuit Court of Washington, D.C., found that both sets of regulations promulgated under CERCLA and the Clean Water Act would have to be revised to make them consistent with the Acts. In particular, the Type-B regulations were found to be overly restrictive in setting out how damages could be assessed.

The efforts of the governments to develop a scientific program to assess the effects of the Exxon spill were shaped and in large part determined by the need to focus on effects that could support assessments of monetary damages. At this point, it is useful to emphasize the distinction between *injuries* to natural resources and *damages* associated with those injuries. Injuries are effects on or changes in the natural resources themselves, for example, the mortality of bivalve mollusks and crabs, dieback of marsh vegetation, or lowered reproductive success among colonial wading birds. Damages are calculated by placing a value on the resources that have been lost or injured or on the benefits (or "services") that they provide to society. The field program developed by the governments was designed to provide data on injuries to natural resources. Other experts and techniques would be used to assess damages.

Initial Actions for Assessment

The five involved governments (the United States; New Jersey; New York State; New York City; and the city of Elizabeth, New Jersey) determined that both coordination and collaboration would be essential to a successful assessment of the spill's effects. During the month following the spill, technical experts from each of the governments worked together to develop a preliminary scope of work for an assessment of the short-term injuries associated with the spill (i.e., those that occurred and would be observable for a year after the spill).

In the days following the spill, government scientists made initial field inspections and began to contact experts familiar with the ecology and resources of the Kill and who had the means to assess oil spill effects. Concurrently, scientists who had worked in the Kill for a number of years made immediate assessments of their study sites and of injuries to their study animals. Just over 2 weeks after the spill, legal and technical staff from all involved agencies met in Trenton to map strategies for handling the legal (both civil and criminal) and technical aspects of their various cases.

Agency staff from the federal government (primarily the National Oceanic and Atmospheric Administration/Damage Assessment Center and National Marine Fisheries Service, with assistance from the Department of the Interior/U.S. Fish and Wildlife Service), New Jersey (Department of Environmental Protection, now Department of Environmental Protection and Energy, coordinated by the Division of Science and Research), New York (Department of Law/Environmental Protection Bureau and Department of Environmental Conservation), New York City (Department of Environmental Protection and Department of Parks and Recreation/Natural Resources Group), and the city of Elizabeth, New Jersey (Department of Health, Welfare and Housing) quickly recognized the need to draft a formal plan for the assessment of the spill's effects (the injury assessment) and for the valuation of the harm attributable to the spill in dollar terms (the damage assessment). Moreover, they realized that even if Exxon were to be required to fund the injury assessment (see "Funding the Study," below), the governments themselves would need to retain quality-assurance responsibility for overseeing all of the scientific and technical work, which would be carried out by a contractor (as opposed to requiring that Exxon carry out a specified scope of work at its own expense).

Thus was born the Governments Technical Committee (GTC), which on January 26, 1990, began reviewing a draft scope of services as prepared by New Jersey's Division of Science and Research and saw the draft through several revisions. The final list of services was completed on February 16 and distributed as a request for proposals to three qualified contractors. Funding considerations led the committee to designate New Jersey's Division of Science and Research as the lead agency for managing the assessment, as described below.

The GTC saw the injury assessment as one part of a comprehensive governmental response to the spill. It adopted a tiered approach, as follows:

> Tier I—Cleanup Action: Cleanup of surface oil slick and, where appropriate, vegetation; continued monitoring of the extent of the spill.

> Tier II—Short-term Action: Mapping, sampling, and analysis of the extent of the spill to contribute immediate information to assessments of injuries to natural resources and of associated monetary damages.

> Tier III—Long-term Action: Long-term studies designed to determine subtle, long-term chronic effects on the ecosystem.

The injury assessment, then, would combine tier II (the data compilation, mapping, sampling, and analytical chemistry tasks described in "Study Design and Results," below) and elements of tier III (the assessments of colonial wading birds and benthic invertebrates, described in the same section).

Funding the Study: The Importance of the Spill Act

Funding for the injury assessment was as difficult an issue for the governments to address as the scope of the assessment was. None of the agencies could easily and quickly locate a separate source of funds from its own budget to pay for the assessment. Direct funding by Exxon would raise the potential problem of undue influence over the study by the spiller. A powerful provision of the New Jersey Spill Compensation and Control Act (the Spill Act), however, allowed the governments to avoid this problem.

The Spill Act establishes a fund to pay for the costs of removing petroleum hydrocarbon discharges, compensating those injured by discharges, and conducting investigations of spill sites. The Department of Environmental Protection and Energy may draw on the fund; alternatively, it can direct a discharger to pay to the department the monies required to fund the necessary work, in this case, the investigation. Companies that fail to comply with a directive to this effect are liable for three times the cost of any work then performed by the department.

The department issued such a "pay-me" directive to Exxon in February 1990. The directive ordered the company to pay the projected cost of the injury assessment ($661,250) within 10 days and included as an attachment the preliminary scope of work that the governments had designed. Exxon complied with the directive, and in March a contract for the work was signed with Louis Berger & Associates of South Orange, New Jersey, the contractor selected by the GTC.

The New Jersey Spill Act thus provided a critical mechanism for funding the initial injury assessment. Information about the scope of the assessment had to be disclosed to Exxon as part of the directive, but the company had no control over the design or conduct of the study, nor did it have access to any of the data while the study was in progress. The availability of a source of funds (either from the fund itself or, as in this case, from the spiller via the directive mechanism) allowed the study to be designed and controlled entirely by the governments.

Limitations and Problems of Damage Assessment

In addition to legal requirements, other factors also affected the potential scope and completeness of the injury assessment. First, the time needed to design the study, issue the directive, select the contractor, and choose sampling locations meant that field sampling could not begin until well after the spill. Ideally, field sampling would take place within days or weeks of the spill to minimize the effects of weathering.

The lack of baseline information on the affected environment is a serious problem in every assessment of oil spill effects. In this respect, however, the situation in the Arthur Kill at the time of the spill was better than most. Scientists had already collected data on colonial wading birds, forage fish, and invertebrates (see "Study Design and Results," below). Nevertheless, few other biological data were available on which an impact assessment could be based, and the overall paucity of data on the Arthur Kill ecosystem made an assessment of changes difficult.

Similarly, the well-documented history of environmental insults suffered by the Arthur Kill and the adjoining creeks and marshes left a less-than-pristine baseline against which to document the effects of the Exxon spill. For example, the elevated background levels of petroleum hydrocarbons required the use of chemical "fingerprinting" techniques to demonstrate that it was the oil from the Exxon spill that had affected a particular site.

Finally, in the months following the Exxon spill, a number of other spills took place in the Arthur Kill and adjacent waters. Among these were the spill of approximately 30,000 gallons of no. 6 oil from the barge *Eklof* on February 28; the explosion and fire on the barge *Cibro Savannah* on March 6, with the release of up to 127,000 gallons of no. 2 oil; the grounding of the tanker *BT Nautilus* in the Kill van Kull on June 7, resulting in the release of some 260,000 gallons of no. 6 oil; and the spill of 40,000 gallons of no. 2 oil from the tanker *Maritrans* on July 18.

Study Design and Results

As mentioned, the injury assessment was carried out through a contract with Louis Berger & Associates (LBA), with the help of several subcontractors. The project consisted of five distinct tasks, described briefly here.

Compilation of Existing Information

This task included a literature search and contacts with area researchers to determine the extent of background information on the affected environment. LBA contacted dozens of researchers, public agencies, and private firms during this effort and identified no fewer than twenty groups that had conducted previous investigations of the Arthur Kill and found eight sets of pre- and postspill samples. LBA also created a preliminary spill extent map, based on field reports from response groups from both states and the federal government. All of this information was used by the GTC in choosing sampling stations for the sediment- and biological-sampling programs.

Mapping

Working with several subcontractors and with staff from the Division of Science and Research's Bureau of Geographical and Statistical Analysis, LBA compiled on digitized maps compatible with the Division of Science and Research's Geographical Information System (GIS) all available information on spill extent and on the degree of oiling. Data layers included shoreline, habitat, and wetland types; land use; property ownership; and sampling locations. The mapped information proved to be of critical importance in helping the governments to determine the potential extent of the spill's effects. Because of their ecological importance and high primary and secondary productivity, intertidal marshes were of particular concern to the governments. Using the GIS, the Division of Science and Research calculated that the spill could have affected 173 acres of intertidal marsh—107 acres in New York and 66 acres in New Jersey—and nearly as many acres of mud flats. The GIS calculations, especially those for intertidal marshes, played an important role in the eventual settlement of the litigation and in decisions of how to allocate between the two states the funds collected under the settlement.

Sediment Sampling

During April and May LBA carried out an extensive sediment-sampling program in the Arthur Kill and adjacent waters, at eleven "intensive" sampling stations (three in areas known on the basis of field reports to have experienced heavy oiling, three in areas of medium oiling, and three in areas of light oiling, with two control stations in areas unaffected by the spill) and twenty-six "additional" stations (thirteen distributed along the New Jersey and New York shorelines, and thirteen located at the mouths and along the lengths of tributaries, including the Elizabeth River, Rahway River, Hackensack River, Morses Creek, Piles Creek, Old Place Creek, Merrills Creek, Sawmill Creek, and Neck Creek (Fig. 2.1). Sediment cores were collected in three regions of the intertidal zone and in the subtidal zone at the intensive sites, and in the medium intertidal or subtidal zone at the additional sites (Fig. 2.2). Cores were sectioned at 2-centimeter intervals to determine the depth of oil penetration. All field sampling and sample preparation activities were performed under strict quality-assurance/quality-control requirements.

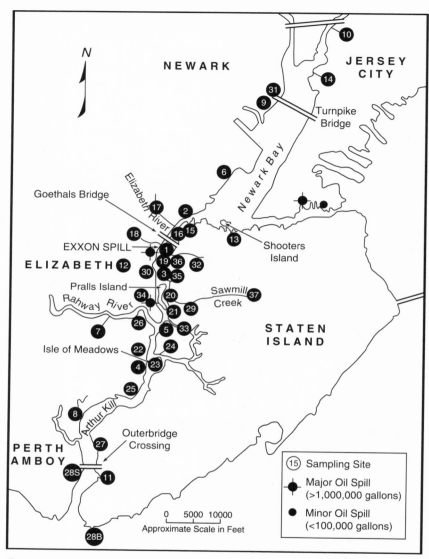

Figure 2.1. Location of sampling stations (after Louis Berger & Associates, Inc., 1991).

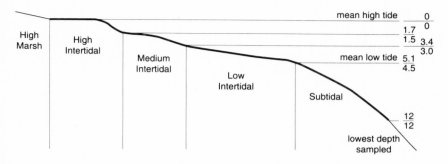

SALT MARSH PROFILE

Figure 2.2. Zones of sediment and benthic sampling (after Louis Berger & Associates, Inc., 1991). Feet below mean high tide (Arthur Kill–Newark Bay/Kill van Kull)

Biological Sampling and Assessments

Biological samples were collected at the same stations where sediment cores were collected. The target organisms were bivalve mollusks, chosen because, as sedentary species, they would reflect each site's specific exposure to oil from the spill. Efforts were made to collect samples of four species (soft-shelled clam, *Mya arenaria;* hard-shelled clam, *Mercenaria mercenaria;* blue mussel, *Mytilus edulis;* and ribbed mussel, *Geukensia demissus*), but only ribbed mussels were collected in significant numbers.

In addition, LBA undertook counts of live and dead ribbed mussels throughout the study area (see Fig. 2.1). Ribbed mussels were found at widely varying densities in the high intertidal zone at thirty-two of the thirty-seven stations; the highest densities occurred at four stations in the southern portion of the study area. The percentage of dead mussels ranged from 0 to 100 percent. the beds with the highest percentages of dead mussels (>67 percent) were located in and around Morses Creek (the location of the Exxon spill), upstream in Neck Creek in Staten Island, and at Tufts Point in Carteret, New Jersey—all areas that were heavily oiled by the Exxon spill (Table 1). Not all heavily oiled areas, however, had high percentages of dead mussels. To some extent, mortality depended on how quickly the oil was removed by tides. Conversely, mussel beds near the site of the March 6 *Cibro Savannah* accident were relatively unaffected, perhaps because relatively little of the oil from that spill reached the shoreline.

LBA also compiled data on the numbers of dead and injured birds that had been collected between the time of the spill and February 22 (Tables 2.2 and 2.3); a total of 654 dead birds were documented, although the

Table 2.1. Results of the Ribbed Mussel Survey for the High Intertidal Zone

| Site name | Site no. | Area sampled (sq. m) | No. of mussels counted | | | Total[a] | Density[b] | % of dead mussels[c] |
			Live	Dead whole shells	Dead half shells			
Goethals (s)	1	5	12	36	52	48	9.6	75.0
Elizabethport Marsh	2	5	46	183	24	229	45.8	79.9
Pralls Island (ne)	3	5	47	3	29	50	10.0	6.0
Opposite Cedar Point	4	5	463	103	7	566	113.2	18.2
Adjacent to Neck Creek	5	5	88	73	37	161	32.2	45.3
Elizabethport (n)	6	5	101	20	12	121	24.2	16.5
Rahway River (w)	7	5	23	15	24	38	7.6	39.5
Adjacent to Smith Creek	8	5	654	80	118	734	146.8	10.9
Turnpike Bridge (s)	9	5	0	0	0	[d]		0.0
Hackensack River	10	5	11	0	7	11	2.2	0.0
Outerbridge (s)	11	5	363	18	36	381	76.2	4.7
Piles Creek (w)	12	5	17	4	13	21	4.2	19.0
Kill van Kull (w)	13	5	53	17	24	70	14.0	24.3
Droyers Point	14					[e]		
Howland	15	5	0	2	4	2	0.4	100.0
Old Place Creek (mouth)	16	5	6	60	7	66	13.2	90.9
Elizabeth River (up)	17	5	0	0	0	[d]		
Morses Creek	18	5	0	1	1	1	0.2	100.0
Gulfport Marsh	19	5	2	49	7	51	10.2	96.1

Site	No.					Total[a]	% Dead[c]	Density[b]
Sawmill Creek (mouth)	20	5	39	2	5	41	8.2	4.9
Pralls Island (s)	21	5	147	2	52	149	29.8	1.3
CITGO	22	3.75	353	72	64	425	113.3	16.9
Isle of Meadows	23	5	153	59	91	212	42.4	27.8
Fresh Kills	24	5	88	98	38	186	37.2	52.7
Tufts Point	25	5	10	29	2	39	7.8	74.4
Rahway River (up)	26	5	199	11	34	210	42.0	5.2
Kreischerville	27	2.75	992	117	62	1,109	403.3	10.6
Sawmill Creek (up)	29	5	20	3	3	23	4.6	13.0
Piles Creek (up)	30	5	0	0	1	d	—	—
Turnpike Bridge (n)	31	5	98	2	1	100	20.0	2.0
Old Place Creek (in)	32	5	162	7	13	169	33.8	4.1
Neck Creek (up)	33	5	60	192	14	252	50.4	76.2
Pralls Island (w)	34	5	62	24	62	86	17.2	27.9
Merrills Creek	35	5	9	5	5	14	2.8	35.7
Old Place Creek (up)	36	5	116	8	14	124	24.8	6.5
Sawmill Creek (in)	37	5	0	0	0	d	—	—

SOURCE: Louis Berger & Associates, Inc., 1991.

NOTE: Results compiled from the biological sampling program conducted between 4/18/90 and 5/14/90.

[a]Total = Live Mussels + Dead Whole Shells.

[b]Density = Total/Area Sampled.

[c]% of Dead Mussels = Whole Shells/Total × 100.

[d]Poor or sparse mussel habitat present at site.

[e]Mussel habitat within the high intertidal zone was not present at site.

Table 2.2. Total Number of Birds Found Dead and Alive by Location (Through February 22, 1990)

Location	Dead birds	Live birds
Sawmill Creek	108	8
Neck Creek	85	8
Pralls Island	88	19
Gulfport	50	2
Isle of Meadows	30	11
Piles Creek	30	9
Staten Island (gen.)	22	7
Bayway Refinery	12	16
ConEd (Stat. Is.)	59	10
Tufts Pt.	12	2
Old Place Creek	29	0
Smith Creek	4	2
Shooters Island	2	3
Port Reading Marsh	5	0
Elizabethport	13	1
GATX Terminal	2	4
Goethals Marsh	1	0
Elizabeth River	1	0
Elizabeth Point Park	0	1
Newark Bay	0	1
Outerbridge Marsh	1	0
Tremley	3	1
Arthur Kill	4	5
Robbins Reef	2	0
Jamaica Bay	3	0
AMAX	2	0
Carteret	2	0
Perth Amboy	1	0
May's Landing	12	0
CITGO	0	1
Rahway River	6	4
Northville	0	1
Rossville	28	8
Pralls Creek	0	1
Unknown	37	25
Total	654	150

SOURCE: Louis Berger & Associates, Inc., 1991.

Table 2.3. Summary of Birds Found Dead
Following the January 1990 Oil Spill

Group	Number	% of Total
Gulls	289	49
Cormorants	4	—
Herons	19	3
Mallards	135	23
Other ducks	148	25
Total	595	100

NOTE: A more detailed list can be found in Table 13.1.

actual toll was probably higher. Of even greater concern to the GTC, however, were the spill's effects on the "harbor herons," the wading birds that had established thriving breeding colonies on Pralls Island, Isle of Meadows, and Shooters Island in the 1970s and had come, in the minds of many, to symbolize the continuing recovery of the Arthur Kill from decades of environmental abuse. Because of the work of Dr. Katharine Parsons of the Manomet (Massachusetts) Bird Observatory and her collaborators, who had been studying the colonies for several years under the Harbor Herons Project, an excellent database for the birds existed, providing a baseline against which the spill's effects on the birds could be determined.

Under a subcontract with LBA, Parsons conducted an assessment of abundance, mortality, reproductive success, and feeding behavior among the harbor herons. Briefly, she found that species that feed in tidal habitats experienced lowered reproductive success and disrupted foraging behavior during the spring 1990 breeding season, whereas omnivorous species and those that feed in upland habitats did not show significant differences as compared with previous years. For a complete discussion of Parsons's work, see Chapter 13.

Drs. Joanna Burger and John Brzorad of Rutgers University conducted assessments of fish abundance in marsh creeks as well as assessments of abundance and behavioral patterns in benthic invertebrates. This provided data on different trophic levels; the fish and invertebrates examined also serve as the prey base for the herons and other colonial wading birds that nest in the Kill. See Chapter 11 for a discussion of this work.

Chemical Analyses

Selected surface sediment and tissue samples were analyzed for total petroleum hydrocarbons; sediments were also analyzed via GC-FID-PID

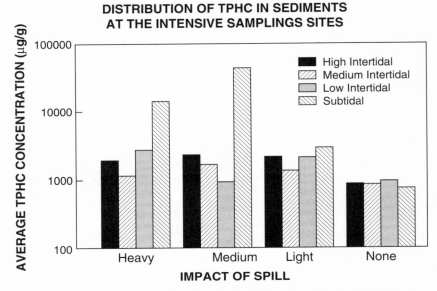

Figure 2.3. Distribution of total petroleum hydrocarbons in sediments at the intensive sampling sites (after Louis Berger & Associates, Inc., 1991).

to allow chemical fingerprinting of the hydrocarbons present in the sample. Fingerprinting is an analytical technique that allows one to compare the chromatographic pattern of an unknown mixture of carbohydrons with the patterns of known source or origin. These results could then be compared with reference samples taken from the Exxon pipeline and with samples from the *Cibro Savannah*. This was necessary because the latter spill occurred before oil was collected for fingerprints. Further analyses via GC-MS were contemplated but carried out only on reference samples.

Sediments from areas affected by the Exxon spill had higher concentrations of total petroleum hydrocarbons than did sediments from control areas (Fig. 2.3 and 2.4). Moreover, the highest levels (up to 120,000 $\mu g/g$) were found in areas that were heavily oiled by the Exxon spill. The total petroleum hydrocarbon content in bivalve tissue samples varied between 60 and 1,400 micrograms/gram, with no correlation with levels in the associated sediments. This puzzling result might actually have been expected. Since tissue samples could be collected only from live organisms and at stations where oiling was heaviest, the stock of live mussels was severely depleted, and the distribution of total petroleum hydrocarbon concentrations in the tissue samples would likely have been heavily

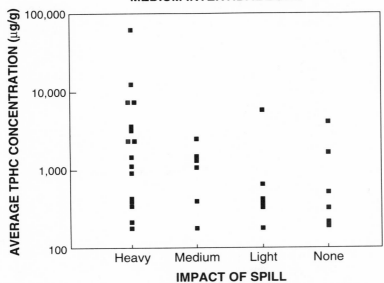

Figure 2.4. Total petroleum hydrocarbons in sediments in the medium intertidal range (after Louis Berger & Associates, Inc., 1991).

skewed downward. Mussels that might have contained the highest levels of total petroleum hydrocarbons probably died before they could be collected.

The chemical fingerprinting used three criteria to determine whether oil in a given sample matched oil in the Exxon reference samples. First, the sample chromatogram was examined qualitatively for the presence of no. 2 fuel oil (i.e., the presence of peaks in the C_{14} to C_{19} range). For those samples that were determined to contain no. 2 fuel oil, a series of ratios of peak areas were calculated for eight alkanes in the "heart-cut" (C_{14} to C_{20}) of the chromatogram: the normal alkanes from nC_{14} to nC_{19}, plus the isoprenoids pristane (2,6,10,14-tetramethylpentadecane, or an isomeric C_{19}) and phytane (2, 6, 10, 14-tetramethylhexadecane, or an isomeric C_{20}). Compounds in this region of the chromatogram were selected because they are more resistant to weathering than are lower-boiling-point compounds and thus would be more representative of the spilled product—a critical factor in the study, since sampling occurred some 3 months after the spill. The peak area for each selected alkane was divided by the peak area for the same compound in the reference sample, and the resulting ratio divided by a similar ratio for nC_{16} in the environmental sample and the reference sample. The nC_{16} thus acted as an internal standard.

Finally, the ratio of the resistant isoprenoids, pristane and phytane, in each sample was calculated and compared with the same ratio in the reference samples.

With this information, a sample was considered to match the Exxon oil if:

1. the sample was determined to contain no. 2 fuel oil;

2. the pristane/phytane ratio in the sample was not more than 10 percent above the ratio for the Exxon oil of 0.97 (the ratio for the *Cibro Savannah* oil was 1.61); and

3. at least five of the seven double ratios were within ±20 percent of the corresponding ratio for the Exxon oil.

Under these criteria, matches to Exxon's pipeline product were found for sediments from eight stations (Table 2.4), all of which were heavily oiled by the Exxon spill, were located within 3 miles of the spill site, and had total petroleum hydrocarbon concentrations in sediments between 2,200 and 120,000 micrograms/gram, considerably above the average for the study area.

The results of LBA's sediment and biological analyses were thus broadly consistent (Table 2.4): considerable overlap existed between stations that had the highest concentrations of total petroleum hydrocarbons, those with the highest percentages of dead ribbed mussels, and those for which matches to Exxon's oil were found.

The Settlement and Beyond

Partly as a result of the work of the GTC, in March 1991 Exxon and the five involved governments reached a settlement in the civil and criminal lawsuits against the company. In the civil-consent order, Exxon agreed to pay $10,000,000 in natural-resource damages. The city of Elizabeth and the city of New York received approximately $250,000 and $200,000, respectively, for spill response costs. The remaining $9,550,000 was split between New Jersey and New York in a 38:62 ratio, reflecting the relative areas of wetlands in each state that had been affected by the spill. The two states were to use these funds to acquire wetlands, to restore natural resources in the area of the spill, and to fund studies of resources in the area. The federal government, as a natural-resource trustee, was to have a voice in decisions on how the funds were spent. In addition, an advisory group consisting of representatives of regional environmental organizations and Exxon was to monitor the governments' progress and provide suggestions.

Under the settlement of the criminal case against Exxon, the company agreed to pay a criminal penalty of $5,000,000. This amount was divided among the governments as follows: Approximately $1,333,000 to the federal government, of which $200,000 served as a penalty under the Clean Water Act; approximately $1,833,000 to New Jersey and New York.

Much of the criminal settlement money was designated by the governments to cover response costs and for environmental enforcement activities. However, more than $1.5 million of the $5 million was set aside as the Arthur Kill Trust Fund, to be administered jointly by the governments to support restoration activities in the area. Thus, of the $15 million paid by Exxon, approximately $11 million was earmarked for restoration, land acquisition, and studies in the affected area, and much of the rest was to be used for enforcement activities.

Lessons from the Exxon Spill

The study carried out under the direct supervision of New Jersey's Division of Science and Research by Louis Berger & Associates and its subcontractors, and overseen by the Governments Technical Committee, provided critical information for the legal negotiations with Exxon. The funding mechanism provided by the New Jersey Spill Act enabled the governments to generate useful scientific information in a timely manner. Of equal if not greater importance to the success of the assessment and the negotiations with Exxon was the governments' ability to shield the conduct of the study from any direct involvement or oversight by the responsible party.

Substantial delays could have been avoided if the governments (particularly New Jersey, which conducted the bidding process and did the actual contracting with LBA) had had access to a list of "preapproved" contractors that could have carried out the study. We understand that the New York State Department of Environmental Conservation is compiling such a list for its own use for spills in New York.

Despite the ultimate success of the process, initial discussions between the representatives of the involved governments did not always proceed smoothly. The governments had to deal with the seemingly inevitable turf issues that often impede cooperation when a number of agencies are involved. Equity issues—for example, which jurisdiction was most damaged and thus should most benefit from any potential settlement—were just below the surface and were apparent from the

Table 2.4. Summary of Findings from the Sampling Programs

Site name	Site #	Spill impact	Highest TPHC (ppm)	TPHC (ppm) in mussels	% of dead mussels	Match to Exxon oil
Goethals (s)	1	High	12,000	150	75	Match
Elizabethport Marsh	2	High	27,000	210	80	Match
Pralls Island (ne)	3	High	1,600	110	6	
Opposite Cedar Point	4	Medium	5,100	160	18	
Adjacent to Neck Creek	5	Medium	120,000	280	45	Match
Elizabethport (n)	6	Medium	920	230	17	
Rahway River (w)	7	Light	6,700	430	40	
Adjacent to Smith Creek	8	Light	520	990	11	
Turnpike Bridge (s)	9	Light	1,300	N/A	N/A	
Hackensack River	10	None	1,500	800	0	
Outerbridge (s)	11	None	1,100	73/67	5	
Piles Creek (w)	12	None	3,700	780	19	
Kill van Kull (w)	13	High	880	1,200	24	
Droyers Point	14	None	190	N/A	N/A	
Howland	15	High	2,200	290	100	Match
Old Place Creek (mouth)	16	High	3,000	140	91	Match
Elizabeth River (up)	17	High	1,400	N/A	N/A	
Morses Creek	18	High	57,000	N/A	100	
Gulfport Marsh	19	High	360	180	96	

20	Sawmill Creek (mouth)	High	1,100	110	5	
21	Pralls Island (s)	High	170	150	1	
22	CITGO	Medium	1,300	660	17	
23	Isle of Meadows	High	430	180	28	
24	Fresh Kills	High	210	330	53	
25	Tufts Point	Medium	370	N/A	74	
26	Rahway River (up)	Medium	2,300	70	5	
27	Kreischerville	Light	290	60	11	
28	Ploughshare	None	450	240	N/A	
29	Sawmill Creek (up)	High	3,400	86	13	Match
30	Piles Creek (up)	Medium	N/A	N/A	N/A	
31	Turnpike Bridge (n)	Light	600	1,400	2	
32	Old Place Creek (in)	High	2,300	68	4	
33	Neck Creek (up)	High	7,200	930	76	Match
34	Pralls Island (w)	High	12,000	130	28	Match
35	Merrills Creek	High	1,400	740	36	
36	Old Place Creek (up)	High	3,400	770	7	
37	Sawmill Creek (in)	Light	180	1,300	N/A	

SOURCE: Louis Berger & Associates, Inc., 1991.
NOTE: N/A = not available.

beginning. There was certainly the apprehension, if not the reality, that Exxon would pursue a "divide-and-conquer" strategy of seeking to settle separately with one or more of the governments. Questions of how best to coordinate the legal issues with technical questions remained even after creation of the separate legal and technical committees. Some of these controversies were overcome simply by the good sense and goodwill of the individuals involved in the process, who recognized from the start that a settlement in the best interest of the public required that the governments act together. The entire process would have been much simpler, however, if arrangements had been in place beforehand for assessing damages from a spill event in the estuarine system. Although the magnitude of the Exxon pipeline spill was unusual, based on past experience such significant oil spills could certainly be anticipated.

Better baseline information, especially on sensitive habitats and on the status of the biological communities in the system, is critically needed for any future spill assessments. The Exxon spill and its aftermath showed that sound scientific information about the status and trends of natural systems is important not just to scientists but to resource managers, regulators, enforcement agencies, and, ultimately, to the general public. In particular, the Harbor Herons Project, data from which were especially critical to the settlement, should continue not only long enough to give us a better idea of the long-term impacts of the Exxon spill but also to provide essential baseline information against which any future changes in the estuary may be gauged. Research to characterize other trophic levels of the system is also needed. Furthermore, some of us hope that the New York–New Jersey Harbor Estuary Program will provide a mechanism for building a more complete understanding of the entire ecosystem, especially with respect to the impacts of toxic substances and other ecological stresses.

Following the spill, Exxon and other companies in the area pledged cooperation in reducing spill risks and argued that the private sector could put its own house in order without the need for additional legislation and regulation by governments. While the substantial decrease in 1990 and 1991 in the number of oil spill incidents indicates that some of their efforts have been successful, plans for the companies to contribute to cooperative mapping of sensitive ecological areas in the harbor have apparently fallen by the wayside. We hope that this is only a temporary setback, not a loss of interest on the part of the companies as a result of the passage of time since the series of major spills in 1990. Such an attitude could lead to a tragic loss of momentum for spill prevention and for providing agencies with the information they need in advance to respond quickly and effectively if a spill does occur.

The governmental response to the Exxon spill taught and reinforced a number of valuable lessons: be prepared before the spill happens, because once a spill has occurred, events move too quickly to accommodate long deliberations; collect baseline data to allow characterization of the affected area and comparisons of pre- and postspill conditions; and establish agreements and arrangements for cooperative action so that each spill does not require the agencies to "reinvent the wheel."

Literature Cited

Louis Berger & Associates, Inc. 1991. Arthur Kill oil discharge study. 2 vols. Final report submitted to New Jersey Department of Environmental Protection.

3. Legal Considerations   *Gordon J. Johnson*

The Court: How does Exxon plead to the charge. Guilty or not guilty?
Exxon Representative: Guilty.[1]

The 1990 Exxon oil spill in the Arthur Kill led to more than just scientific studies and remedial efforts to address the spill's ecological impacts. Government lawyers and investigators responded along with the biologists and chemists, not to mention the legal and scientific team promptly assembled by the corporation itself. This chapter describes the legal context of the Arthur Kill spill, particularly the governments' claims for "natural resource damages," the interplay between legal claims and scientific inquiry, and the ultimate settlement of the civil and criminal claims against the company. Finally, revisions to the federal law governing oil spills, which were enacted in August 1990 largely in response to the *Exxon Valdez* and the rash of other spills in 1989 and 1990, are discussed.

Legal Context

Civil Liability

Recent law governing oil spills is an evolution of the tension between the centrality of oil to the U.S. economy and the risks posed by that very fact (Table 3.1). The controlled and unintended release of oil and its byproducts into the environment has been the subject of a welter of laws and regulations, although not until 1990 was a comprehensive federal statute enacted in an effort to require effective, albeit long-term, prophylactic measures, subject violators to severe penalties, create a meaningful cleanup fund, and stiffen liability provisions.

At the time of the Arthur Kill spill, the central statute governing liability of persons who spill oil on or into navigable waters of the United States was Section 311 of the Clean Water Act.[2] As discussed toward the end of the chapter, major provisions of that statute were replaced by the Oil Pollution Act of 1990, which applies to spills on navigable waters, onto adjoining shorelines, or in the "exclusive economic zone," which extends 200 miles offshore. This act covers spills occurring only after its effective date, August 18, 1990, however.

Initially enacted in 1970 in reaction to the Santa Barbara offshore platform discharge,[3] Section 311 provides that the United States may clean up or arrange for the cleanup of spilled oil unless the government

Table 3.1. Statutes Relating to Oil Spills

	Oil spills before 8/19/90	Oil spills after 8/19/90
Statute		
Section 311 of the Clean Water Act, 33 U.S.C. § 1321 (1990)[a]	X	
Oil Pollution Act of 1990, 33 U.S.C. §§ 2701– 2761 (1990)		X
N.Y. Navigation Law, §§ 170 *et seq.*	X	X
New Jersey Spill Compensation and Control Act, N.J.S.A. 58:10–23.11, *et seq.*	X	X
Regulations		
Type-A rule for natural-resource damage assessment, 43 C.F.R. Part 11	X	
Type-B rule for natural-resource damage assessment, 43 C.F.R. Part 11.	X	
NOAA regulations for natural-resource damage assessments for oil spills (unpromulgated as of January 1994)		X

[a]As modified by the Comprehensive Environmental Response, Compensation, and Liability Act of 1980, 42 U.S.C. §§ 9601 *et seq.*

determines that the cleanup will be done properly by the owner or operator. The owners and operators of vessels and facilities that release petroleum on or into the waters of the United States are liable for these "removal costs."[4] Because many spillers, particularly major oil companies and members of what have become regional cleanup consortia, concluded that they can respond and clean up more quickly and at substantially less cost than the government can, and because penalty amounts are determined in part by the actions of a spiller in responding to a spill, spillers usually choose to conduct the cleanup operations themselves under U.S. Coast Guard supervision. Not to be underestimated as an incentive to conduct the cleanup oneself is the negative publicity that may result if the spiller does not respond.

Consequently, after some initial hesitation following the Arthur Kill spill, Exxon began mobilizing cleanup crews and equipment. Under Coast Guard supervision, the long process of capturing and removing floating oil, cleaning fouled shorelines, and taking other measures deemed necessary by the Coast Guard to respond to the oil contamination in the harbor was begun.

Section 311 was amended in 1977 to authorize states and the federal government to recover "the costs of replacing or restoring" damaged natural resources, including the cost of "minimiz[ing] or mitigat[ing] damage to the public health or welfare, including, but not limited to, fish, shellfish, wildlife, and public and private property, shorelines, and beaches."[5] With few exceptions,[6] however, natural resource damages were not sought until the passage of the Comprehensive Environmental Response, Compensation, and Liability Act of 1980, or CERCLA,[7] which also established the so-called Superfund to clean up toxic-substance releases. Even after CERCLA's passage, natural resource damages were only rarely sought. Several reasons have been given for this failure, including lack of money to prosecute suits, absence of accepted methods to calculate damages, and lack of interest in utilizing the remedy, particularly on the part of the federal government (see Olson 1989).

CERCLA excluded from its terms petroleum and its refined fractions, even though they contain chemicals, such as benzene, that are "hazardous substances" covered by CERCLA.[8] "Hazardous substances" refers to a large number of chemicals designated as pollutants, toxic pollutants, or hazardous wastes by other federal statutes or by another provision of CERCLA.[9] The list of such substances is published as 40 C.F.R. Part 302 and includes common organic industrial chemicals, most chlorinated solvents, pesticides, air pollutants, polychlorinated biphenyls, corrosives, and heavy metals. As a consequence, only petroleum spills that happen to be contaminated with "hazardous substances" as a result of the use of the petroleum or of mixing after it has left the refinery can be cleaned up using Superfund monies.[10]

CERCLA, however, clarified the scope of Section 311's natural resource damages provision for oil spills by expanding the remedy and modifying a traditional rule governing its proof in lawsuits for both hazardous substance and oil spill claims. As a result of CERCLA's modifications, an "authorized representative"[11] of the federal or state government could seek natural resource damages, usually acting through a "trustee of natural resources," in Section 311 cases.[12] Compensation could be sought for injuries to or loss of "land, fish, wildlife, biota, air, water, ground water, drinking water supplies, and other such resources."[13]

The resource must be one "belonging to, managed by, held in trust by, appertaining to, or otherwise controlled" by the government.[14] In general, the government holds numerous interests in property in trust for the public, such as the rights of navigation on surface waters, fishing, hunting, beach use, and others, even with respect to private property.[15] While these property interests do not constitute ownership of, for instance, the natural resource of land, they can be diminished in the event of an oil spill. Thus, actual ownership of the damaged resource is not required, and "a substantial degree of government regulation, management or other form of control over property would be sufficient" to make the resource subject to a government damage claim.[16]

This right to damages is distinguishable from the right to reimbursement for the costs of cleanup undertaken by the Coast Guard or by another federal or state agency to remove spilled oil and to clean contaminated shorelines. Removal actions and cleanup efforts *abate* the problem and protect human health and the environment from future harm. Natural resource damages *compensate* the public for injuries to natural resources until abatement actions are completed and for residual harm remaining after abatement. "Customarily, natural resource damages are viewed as the difference between the natural resource in its pristine condition and the natural resource after the cleanup, together with the lost use value and the costs of assessment."[17] This approach recognizes that injury may continue until the abatement process is finished, and in any event abatement activities may not address all of the injuries to natural resources and that further measures to restore or replace injured natural resources may be necessary to make the public whole.

Fortunately, and perhaps most significant from a strict environmental point of view, in CERCLA Congress retained Section 311's requirement that natural resource damages be spent only to "restore, rehabilitate or acquire the equivalent of [the damaged or destroyed] natural resources."[18] Consistent with the concept of trusteeship, the trustee may spend collected damages only on the natural resources under the trustee's control.[19]

Section 311's standard of "strict, joint and several liability" was adopted by CERCLA for hazardous substance spills.[20] In a Section 311 lawsuit, a government seeking recovery need demonstrate only that the named defendant was an owner or operator of the vessel or facility spilling the oil. Because the standard of liability is "strict," the government does not have to demonstrate any fault on the part of the owner or operator. "Joint and several" means that any liable person is responsible for all response costs, not just the portion attributable to that person's share of oil or responsibility when more than one person is responsible for the spill. Defenses to liability in Section 311, which must be proved by the defendant, were limited to "acts of God," that is, situations in which the release was entirely a result of a natural disaster for which preventative measures were not possible or feasible, acts of war, and acts of third parties or negligence of the United States if those acts were the sole cause of the release.[21]

Recognizing that proof of natural resource damages is difficult and disfavored by the common law, Congress provided in CERCLA that "natural resource trustees," appointed by the president and governors of each state, could perform a natural resource damage assessment following both oil and hazardous substance spills. Such an assessment, the costs of which are recoverable if reasonable, would be granted a "rebuttable presumption" in proceedings seeking to recover such damages in cases involving either the release of oil or the release of hazardous substances into the environment, provided that the assessment was conducted in accordance with regulations to be promulgated by the United States.[22]

The effect of the "rebuttable presumption" in actual litigation is unclear. An assessment entitled to that presumption must be disproved by the defendant, and it is otherwise sufficient to establish a trustee's prima facie case. The kind or degree of proof needed to disprove it and whether a trustee would have to support his or her determination with more than just the assessment once the result was attacked have not been determined in actual litigation. Nevertheless, the mere existence of the presumption for assessments conducted according to the regulation is a potent weapon for the government in negotiations between it and a spiller. Most civil cases do not go to trial; instead, they are settled out of court.

In CERCLA, Congress required the promulgation of rules that would allow two alternative ways of assessing natural resource damages in both oil and hazardous substance spills: the "Type-A rule," consisting of "standard procedures for simplified assessments requiring minimal field observations," and the "Type-B rule," consisting of "alternative protocols for performing assessments in individual cases" expected to be used when the spill is more complicated, large, or more damaging.[23] Although Congress directed that the natural resource damage assessment rules be promulgated as regulations within 2 years following CERCLA's passage in December 1980, the federal agency assigned the task, the U.S. Department of the Interior (DOI), did not do so until March 1987.[24]

As promulgated, the DOI's regulations authorize trustees to perform a Type-A simplified assessment only for spills of oil or hazardous substances into marine waters and only by using a DOI-developed computer model that, with minimal inputs such as the size of the spill, wind direction, spill substance, and other data easily gathered when a spill occurs, calculates likely biological and other environmental injuries and specifies a dollar damage figure.[25] Alternatively, the regulations permit a trustee to perform a Type-B assessment for marine spills and all other releases by following more complicated provisions allowing use of various methodologies to determine actual injuries to natural resources and various economic methodologies to calculate, in specified situations, the value of certain "lost uses" of such injured or destroyed resources.[26]

Lost use, or the utility that humans derive from a resource, may be consumptive or nonconsumptive. Typical consumptive uses of natural resources include harvesting of fish, shellfish, and game birds; irrigation of crops; and drinking of groundwater. Nonconsumptive uses, that is, uses that do not consume the resource, include recreational use of parks, bodies of water, and shorelines; birdwatching; hiking; and viewing of beautiful vistas. In addition, natural resources also have option value, the utility that humans place on the opportunity to use a resource in various ways in the future, and existence value, that which humans give merely because the resource exists and that existence brings pleasure. These latter values are also called "passive use" or "nonuse" values.

Rather than applauding the regulations, state trustees uniformly denounced them. Nine states, including both New York and New Jersey, challenged the Type-B regulations immediately and, with respect to the significant issues, successfully. The case, entitled *State of Ohio, et al. v. United States Department of Interior*,[27] (*Ohio*) was brought in the U.S. Court of Appeals for the District of Columbia Circuit and was decided in July 1989. The DOI was ordered to repromulgate the regulations, to emphasize the collection of sufficient damages to allow trustees to restore or replace natural resources injured or destroyed by a spill and to permit trustees to use whatever economic methodologies would most fully capture the true and complete value of lost use and nonuse values of natural resources. The Type-A rule was also challenged and the DOI was ordered by the same court to rewrite it to correct similar deficiencies.[28] The court decisions also strongly endorsed and clarified the nature and scope of the natural-resource damage remedy, setting the stage for easier utilization of that cause of action by states and the federal government.

The breadth of recoverable natural resource damages is evident upon review of the unchallenged portion of the DOI regulations defining injury to resources, the court's ruling in *Ohio*, and other courts' opinions construing the remedy after the *Ohio* decision. Injury is defined in the regulations as "a measurable adverse change, either long- or short-term, in the chemical or physical quality or the viability of a natural resource."[29] Proof of injury is established by evidence of such change resulting directly or indirectly from exposure to the released oil or the products of its degradation. The regulations list methodologies and tests used to determine injury, and upon demonstration by the trustee that the test results meet "acceptance criteria," such as results above certain contamination limits in two properly collected samples, injury is deemed proven.[30]

For groundwater, surface water, air, and geological (soil, sediments, etc.) resources, contamination in excess of a statutory or regulatory limit constitutes a legally recognized injury. For biological resources, the regulations specify a series of "biological responses" (e.g., death, disease, reproductive failure, growth abnormalities, uncharacteristic tumors) and tests to determine and measure those responses. Upon "establishment of a statistically significant difference in the biological response between samples from populations in the assessment area and in the control area," these responses constitute a demonstration of injury because the DOI has already concluded that such responses satisfy the biological injury acceptance criteria. Other biological responses may also demonstrate injury if they meet all of the criteria, which for biological injury include demonstration that the response is often the result of exposure to oil, that exposure to oil is known to cause the response in free-ranging organisms and in controlled experiments, and that the biological response measurement is practical to perform and produces scientifically valid results.[31] Contamination above action or tolerance levels governing consumption of biota (e.g., fish), such as Food and Drug Administration tolerance limits, also constitutes injury.[32]

"Aesthetic injuries," which reduce one's enjoyment of a natural park or natural vista, are also compensable: "The term 'aesthetic injury' . . . describe[s] pollution that mars the physical appearance of a natural resource, or in some other manner offends the senses, for example, by a foul stench. Also included in that term is pollution that contaminates a natural resource or otherwise renders it less than pristine or less than it was prior to release."[33]

The *Ohio* court held that compensation must be paid for the loss of any portion of the full value of the natural resource, not merely that portion measured in the market and revealed by market pricing. "From the bald eagle to the blue whale and snail darter, natural resources have values that are not fully captured by the market system."[34] Consequently, the court threw out the regulation's provisions limiting proof of the value of a resource solely to its market value if the resource was traded in a competitive market. As examples, the court explained that the value of seals could not be limited to the price of their pelts, about $15 per animal, or the value of a national park limited to the sum of entrance fees paid by the public, as originally intended by the regulations.

Thus, a trustee could recover damages for the loss or diminution of a natural resource's full value, including its option and existence values as well as traditional use values.[35] Moreover, the spiller also has to pay for the restoration of injured resources or their replacement through the acquisition of equivalent resources unless the restoration or replacement costs are grossly disproportionate to the resources' value.[36]

In April 1991, the DOI proposed, but as of January 1994 had not yet finalized, regulatory changes to conform to the *Ohio* court's decision. The department also stated that it will revise the Type-A model and expand it to cover spills in the Great Lakes.[37]

The *Ohio* decision seemed to help change the federal government's attitude toward natural-resource damages. Formerly hostile and desirous of limiting the scope of damages, federal officials became interested in utilizing the cause of action to restore damaged resources. Undoubtedly, Exxon's *Valdez* spill was a major contributor to this change in attitude. The *Ohio* decision also eliminated some of the roadblocks to easy utilization of the remedy by state trustees.

Thus, when the Arthur Kill spill occurred, the federal government, New York, and New Jersey recognized that they should begin the process of recovering damages for natural resource injuries under federal as well as state law. While not as developed as federal law, New York and New Jersey state laws also provide for recovery of natural resource damages.[38] Remarkably, after the first few days of frantic government activity following the spill, attorneys from the offices of the U.S. Department of Justice; the New York State and New Jersey Attorneys General; New York City Corporation Counsel; scientific staff and attorneys of the federal and state agencies with authority over natural resources, including scientists from the New York Attorney General's office; and officials from the city of Elizabeth (whose shoreline and marina had been heavily oiled) began to meet to discuss legal options. The five governments quickly concluded that a Type-A assessment would not adequately capture the probable scope of injury or calculate damages properly, given the apparent impact of this spill. Thus, the immediate legal and strategic need was to gather scientific data: The governments needed to begin the studies necessary for a Type-B damage assessment that would hold up in court later (see Chapter 2).

Although any competent scientist could come up with a huge list of different studies that could be undertaken regarding the spill's effects, the governments' legal team placed priority on determining compensable injuries to resources while and where evidence was more readily available. The governments wanted to begin studies to chronicle pre-spill baseline conditions, determine and map oil flow, and record transitory and initial biological impacts on birds and certain benthic species (see Chapters 7–14). However, while a study of one particular species of microorganism or adult non-bottom-dwelling fish might yield interesting, indeed valuable, information on the effects of oil on those creatures (particularly for a scientist already studying that creature), such studies were of low priority because the information gathered would not likely increase an eventual dollar recovery from Exxon. Given the absence of readily available funds to the governments, the limitation on recovery of assessment costs to only "reasonable" costs (limited by the DOI to no more than the anticipated damage amount),[39] the desire to induce Exxon's compliance with any demand to pay the costs, and the desire to collect data likely to maximize recovery but not data that might induce a judge or jury to limit recovery, careful thought was given to the type of studies that should be conducted following the spill.

The governments met with Exxon's in-house and outside legal counsel and its managers to discuss matters that either side thought might be resolved without litigation. Given that civil liability for damages was, for all practical purposes, incontestable because fault was irrelevant to liability and because Exxon could not shift the liability to a third party by proving that that party was solely responsible for the spill, Exxon publicly announced that it would pay all appropriate damages. The problem was to determine what was appropriate. By early March 1990, the governments had prepared a description of the scientific data collection and analyses they concluded were necessary to begin a Type-B natural-resource damage assessment, and asked Exxon to fund an independent consultant to perform the initial assessment. Unable to obtain a quick commitment to pay the costs of these studies, the governments joined in a New Jersey directive requiring Exxon to fund the initial assessment, at a cost of $661,250. The governments relied upon a provision of the New Jersey Spill Compensation and Control Act,[40] which subjects a party failing to comply with a directive to triple the cost of the damages. Exxon paid, and the work began.

Criminal Liability

Immediately after the spill, criminal investigators and prosecutors from the offices of the New York and New Jersey Attorneys General and that of the Staten Island (Richmond County) District Attorney began their investigations of possible nonfederal criminal charges arising out of the spill. Because the spill had entered New York State waters, New York had jurisdiction over any criminal activity leading to the spill. New Jersey, the site of the Exxon plant and the release of oil itself, also had obvious jurisdiction. Grand juries were convened, experts hired, and a determination of the events behind the spill began in each jurisdiction. Both the U.S. Attorney for the District of New Jersey and the U.S. Attorney for the Eastern District of New York, the federal judicial district covering the harbor waters where the spill occurred and the New York land onto which approximately two-thirds of the oil spread, had jurisdiction to prosecute federal criminal charges arising out of the spill. They agreed that any federal criminal charges would be prosecuted in the federal court for the New Jersey District, and the United States' civil liability claims would be pursued in the federal court for the eastern district of New York. Consequently, the U.S. Attorney for New Jersey soon began his own criminal investigation, in close cooperation with the two states.

In contrast to the civil liability for natural-resource damages, criminal charges had to be supported by a criminal mental state: The charging government had to demonstrate that Exxon had acted, or failed to take required action, intentionally or through criminal negligence.[41] While there are in the environmental area a limited number of federal "strict liability" criminal statutes for which demonstration of criminal intent is not necessary, such as the statute criminalizing the killing of migratory birds at certain times or in a prohibited manner,[42] criminal prosecutors believed that more serious charges requiring a demonstration of criminal intent were appropriate given the circumstances leading up to the spill.

To avoid legal ethical conflicts that can arise when a government is conducting both civil and criminal investigations and prosecutions, the governments' civil attorneys and criminal prosecutors guided their investigations and planned their strategies separately. Resolution of potential criminal charges was discussed by government prosecutors separately from those concerning civil liability, although any resolution was likely to be interdependent. (Exxon, of course, did not have to split its discussions, although the company retained lawyers experienced in criminal law to engage in those discussions with the government prosecutors.) Eventually, in the fall of 1990, as the initial results of the damage assessment became known, discussions again began in earnest between Exxon attorneys and managers and government lawyers and staff, civil and criminal. Exxon indeed desired a "global" settlement, one that would resolve civil and criminal liability in all jurisdictions. This settlement was reached in March 1991.

The Criminal and Civil Settlements

Potential criminal charges of the three jurisdictions that could possibly bring them—the United States, New York State, and New Jersey—were the subject of a $5,000,000 plea bargain. Exxon agreed to plead guilty to the federal crime of negligent discharge of a pollutant, 567,000 gallons of heating oil, into the Arthur Kill, a navigable water of the United States, in the federal district court in New Jersey.[43]

Exxon admitted that plant personnel overrode automatic shutdowns of the interrefinery pipeline (IRPL) on the night of January 1–2, 1990, and that Exxon had negligently failed to train and supervise its staff. The maximum federal fine of $200,000 was imposed, together with a mandatory assessment of $125 payable to the federal crime victims fund. Exxon was placed on probation and ordered, as a condition of probation, to pay criminal restitution, that is, compensation for damages to resources injured by the spill and reimbursement of government costs, totaling $4,799,875. The probation lasted only until Exxon made the payments, which were made the same day the plea was taken and the sentence imposed. The three governments agreed not to bring any further criminal charges arising out of the incident against Exxon or its officers and employees.

Part of the $5,000,000 was distributed to the three governments, and the remainder was placed in a trust fund administered by the court. The criminal fine of $200,000 was paid to the federal treasury and the $125 payment to the crime victims fund. New York and New Jersey each were earmarked $1,833,333 of the restitution. Of New York's share, $1,449,745 was paid to two New York State accounts used by its Department of Environmental Conservation to investigate and enforce environmental laws and conduct natural resource damage assessments and investigations, and $26,255 was used to reimburse laboratory costs. All but $100,000 of New Jersey's share was also paid into similar enforcement accounts and to reimburse the state for various costs associated with the prosecution and cleanup. The remainder of the payments to the states, $357,333 of New York's and $100,000 of New Jersey's, was deposited into the Court Registry Arthur Kill Trust Fund in the U.S. District Court for the District of New Jersey, as was the bulk of the federal portion, $1,105,247, establishing a trust fund of $1,562,580. The remainder of the total federal portion, $27,962, reimbursed various federal agencies involved in natural resource damage assessment for costs incurred as a result of the spill.

The civil settlement, filed in the U.S. District Court for the Eastern District of New York in Brooklyn,[44] directly addressed the federal and states' claims for natural resource damages and the need for preventative measures. Exxon agreed to pay the three governments and the cities of Elizabeth and New York a total of $10,000,000 over 5 years, with the bulk of the payments—$7,500,000—to be paid within 2 years after announcement of the settlement on March 20, 1991. Portions of the payments were to be made directly to the two cities, to reimburse them for costs they incurred in cleaning up the spill and for property damage. Also, $300,000 was to be paid directly to the city of Elizabeth to be used to improve and restore its waterfront park. The remainder, $9,300,416, was placed in a trust fund administered by the court, to be used solely for the acquisition of title or conservation easements on land and wetlands within the New York–New Jersey–Arthur Kill harbor and related ecosystems to restore or replace natural resources that had been damaged as a result of the spill; the restoration (including the establishment, maintenance, and provision of public access) and protection of acquired and adjacent lands and wetlands; and the performance of studies in the New York–New Jersey harbor and related ecosystems. No civil penalties were imposed; the punitive aspects of the spill were addressed by the criminal settlement.

The civil settlement also required Exxon to take specified preventative measures, train people to deal with the effects of oil spills and their prevention, and release information. First, the IRPL could not be reopened until Exxon took specified actions and/or complied with orders issued by the U.S. Department of Transportation (DOT), which regulates interstate pipelines. The actions included the provision to the governments of a study regarding the IRPL; written description of pertinent operating characteristics, including the leak detection system; a tour of the IRPL facilities; and a notice designating a knowledgeable Exxon employee as a contact for government inquiries concerning the IRPL. Changes in major operating characteristics of the IRPL also had to be communicated to the governments after the IRPL was reopened. New operating procedures, formal training programs, repair or replacement of the leak detection system, measures to reduce future external damage to the IRPL, and a pressure test would have been required to be developed and implemented before the IRPL was reopened absent a DOT order otherwise governing the IRPL's operation. The IRPL has not been reopened as of January 1994. However, the DOT issued such an order on April 8, 1991, essentially incorporating these elements.

Second, Exxon agreed to train persons designated by the governments concerning marine terminal operations in the New York harbor and to train twenty-five people annually for 5 years in the care and handling of waterfowl exposed to petroleum spills. Finally, Exxon—and the governments—agreed to disclose all data they had collected or developed, subject to specified and narrow legal privilege claims, to promote and advance scientific study of the spill, the Arthur Kill/harbor area, and petroleum spills in general.

The governments required a public comment period before submitting the settlement to the court for approval. A coalition of environmental groups, the Safe Harbor Coalition, criticized the settlement, asserting that the sums to be paid by Exxon were inadequate and that further studies were needed before a settlement could be negotiated.[45] The governments rejected these criticisms, explaining that the data collected indicated that approximately 25 acres of wetlands suffered severe oiling and that the sums collected were more than sufficient to acquire equivalent acreage. In addition, the damages collected through the settlement were at or above the amounts collected in other recent, comparable major spills. More crucial, the governments argued that because the harbor area had been and probably would continue to be subject to petroleum spills after the Exxon spill, injuries discovered after years of scientific studies would be difficult to connect to the Exxon spill.

The governments also noted that an actual trial of claims often entails considerable risk, especially in a new area of the law. The governments' claims of injuries to wetlands, benthic organisms, birds, and other biota, and the manner in which the dollar values were calculated, could be rejected by a judge or jury on grounds that a study was flawed or the proof or methodology inadmissible or inappropriate for legal reasons, or because the harbor was found by a jury or judge, albeit inaccurately, to have been significantly degraded before the spill. Since the negotiations were at arm's length and were adversarial, the governments argued that their decision to settle should be given deference by the court.[46]

The coalition also asked that more complete disclosure of the studies be made before the settlement was approved, that better and more enforceable preventative measures be required, and that the settlement be reopened in the future if damages proved to be greater than anticipated.[47] The governments responded to these criticisms, stating that sufficient data had been released and that further delay in entry would not likely result in any higher recovery; that the preventative measures were enforceable in practice and were probably more than the law would allow a court to impose if the case went to trial; and that given the history of spills in the area, it would not be possible to show in the future that Exxon's spill, rather than later spills, caused unanticipated damage discovered later.[48]

Several other elected officials commented, primarily asking for greater penalties, to which the governments replied that the penalties collected were substantial and constituted a sufficient deterrent. The governments noted that Exxon suspended its tanker and barge operations along the Arthur Kill and Kill van Kull for a period of time to comprehensively evaluate the operations, and consequently had developed and began implementing measures to improve the operations at costs likely to exceed $25,000,000; that an update of these measures would be reported to the governments 1 year following entry of the proposed order; and that Exxon had already paid the governments $661,250 to conduct their own studies of the spill and spent approximately $18,000,000 for other cleanup measures supervised by the Coast Guard.

After review of the submitted papers U.S. District Court judge Edward Korman approved the settlement without any delay or oral presentation.

Oil Pollution Act of 1990

Until the enactment of the Oil Pollution Act of 1990 (OPA),[49] major revisions and consolidations of federal laws governing oil spill liability had died in one or the other houses of Congress over the previous 14 years. The recurring, large spills of 1989 and 1990, however, changed congressional attitudes and made oil spill law reform a priority. The *Valdez* spill occurred in March 1989 (260,000 barrels—about 11 million gallons). It was followed by the 290,000-gallon *World Prodigy* oil spill in Narragansett Bay on June 23, 1989, and on the next day the 250,000-gallon Houston Ship Channel spill and the 300,000-gallon *Presidente Rivera* spill in the Delaware River near the New Jersey–Delaware border. After the Exxon Bayway spill, numerous smaller spills occurred in the Arthur Kill/Kill van Kull, culminating in the 260,000-gallon *BT Nautilus* spill on June 7, 1990. Comprehensive legislation, in the form of OPA, was enacted in August 1990.

In OPA, Congress reshaped the law originally stated by Section 311. OPA made recovery of damages and response costs easier, imposed substantially higher civil penalties for spilling oil and increased criminal penalties, increased the oil spill trust fund so that funds would be available immediately to clean up spills, and revised spill prevention requirements for vessels and facilities. Pertinent here are the provisions that govern oil spill liability.

While Section 311 still pertains to spills occurring before OPA's passage, the liability of spillers for response costs and damages resulting from subsequent spills is now governed by OPA §§ 1001–1019.[50] Congress retained the strict joint and several standard of liability of Section 311 but limited the defenses available to parallel those allowed by CERCLA. While CERCLA's liability standard is the same—strict, joint, and several—the allowed defenses are even more limited than those allowed by Section 311. In OPA, a spiller can avoid liability for releases caused solely by an act of God, as in Section 311, but the "third-party defense" is limited to situations in which the defendant proves that the spill was caused solely by third-party acts that did not "occur[ ] in connection with any contractual relationship with the responsible party."[51] Even if there is no relationship, the owner or operator must demonstrate that it exercised due care and took precautions against foreseeable acts or omissions of the third party.[52] In addition, negligence of the United States as the sole cause of a release no longer constitutes a defense to an owner or operator's liability.

Congress did not, however, expand the class of liable parties in OPA to parallel CERCLA's liability scheme for hazardous substances, although the House of Representatives' version of OPA that went to conference with the Senate would have imposed "secondary liability" on cargo owners.[53] In CERCLA, liability extends not only to the owner or operator of the vessel or facility from which the hazardous substance was released but also to the generator of the substance, that is, the person who arranged for the disposal of the chemical, absent demonstration of one of the limited defenses.[54] Because the generator almost always had a direct or indirect contractual relationship with the disposal site or transporter of a substance, generators can rarely avoid liability for release of chemicals that left their own plants for anticipated lawful, safe disposal even when the generator could not have anticipated that the disposal-site operator might release the substance to the environment. In CERCLA, transporters are also liable for spills at a disposal site when they chose the facility from which there is a release. Thus, OPA liability does not extend as far as CERCLA liability.

OPA allows the necessary damages as well. In addition to state and federal trustees of natural resources, "Indian tribe" and "foreign government" trustees may recover for "injury to, destruction of, loss of, or loss of use of, natural resources," and any recovery for such damages must be spent on the "restoration, rehabilitation, replacement, or acquisition of the equivalent of the natural resource" under trusteeship.[55] Also incorporated into OPA were the expansive definitions of CERCLA regarding natural resources.[56] OPA also allows federal, state, and local governments to recover the net costs of providing increased public services resulting from an oil discharge, lost fees, taxes, and related revenues.[57] Damages for loss of subsistence use of natural resources can be recovered by any claimant "who so uses such natural resources which have been injured, destroyed or lost, without regard to the ownership or management of the resources."[58] Any claimant can recover for personal and real-property damages and loss of profits or impairment of earnings resulting from a spill.[59] Like Section 311, dollar limits on liability are imposed but at significantly higher levels. The limitations are voided and liability is unlimited if the incident is caused by gross negligence, willful misconduct, or the violation of an applicable federal safety, construction, or operating regulation by a responsible party or its agents.[60]

Trustees are responsible for assessing damages and developing and implementing a plan for "the restoration, rehabilitation, replacement, or acquisition of the equivalent, of the natural resource under their trusteeship."[61] The measure of damages is explicit:

> The measure of natural resource damages . . . [is] (A) the cost of restoring, rehabilitating, replacing, or acquiring the equivalent of, the damaged natural resources; (B) the diminution in value of those natural resources pending restoration; plus (C) the reasonable cost of assessing those damages.[62]

The joint House–Senate committee that reconciled the two houses' bills reported that Congress intended that the holdings of the *Ohio* court regarding the measure of damages be followed by the courts in interpreting OPA and when promulgating assessment regulations. The committee also exhorted trustees to work together by coordinating their assessments and the development of restoration plans, and to exercise "joint management or control of the shared resources," noting that one class of trustees cannot preempt the trustee responsibilities of other trustees.[63]

The task of promulgating new regulations to assess damages was expressly given to the Under Secretary of Commerce for Oceans and Atmosphere, part of the National Oceanic and Atmospheric Administration (NOAA), in consultation with the Environmental Protection Agency, the U.S. Fish and Wildlife Service, and the heads of other affected agencies. As before, an assessment is entitled to a rebuttable presumption.[64] NOAA has begun that process, issuing a number of "notices of advance rulemaking" summarizing issues and probable revisions of the DOI-mandated procedures.[65] Although OPA required promulgation of new regulations by August 1992,[66] proposed regulations were not published until January 7, 1994.[67]

Conclusions

Legal liability for owners and operators of vessels or facilities from which petroleum is released is strict, and absent narrow defenses owners and operators are responsible for the costs of cleanup and for natural resource damages regardless of fault. The measure of damages is the sum of the costs of restoring or replacing the damaged or destroyed resources plus the value of the lost uses pending the resources' restoration or recovery. When collected, these sums must be spent on the resources. The 1990 revision of federal oil spill law (the Oil Pollution Act) has expanded the remedies, allowing persons who depend on the damaged resources for income or subsistence to recover damages as well. With the promulgation of regulations that should allow natural resource trustees to more easily determine, compute, and collect damages, governments are more likely in the future to use the law to fund the study and restoration of natural resources after oil spills.

Literature Cited

Olson, E. 1989 (December). *Natural Resource Damages in the Wake of the Ohio and Colorado Decisions: Where Do We Go from Here?* 19 ELR 10551, 10551-2.

Notes

The opinions expressed in this chapter are not necessarily those of New York State or any of its agencies, departments, or officials.

1. Transcript of criminal proceeding, *United States v. Exxon Corp.*, Crim. No. 91-131 (D.N.J. March 20, 1991) at pp. 17–18. (Throughout, specific references to statutes, regulations, and case law are given footnote numbers and appear in this section.)

2. 33 U.S.C. § 1321 (1990).

3. Section 11 of the Federal Water Pollution Control Act, as added by the Water and Environmental Quality Improvement Act of 1970, Pub.L. 91-224, 91st Cong., 2d Sess. (1970). It was renumbered and reenacted in 1972 as Section 311 in substantially the same form, Pub.L. 92-500, 92nd Cong., 2d Sess. (1972).

4. 33 U.S.C. § 1321(f) (1989); see also National Contingency Plan for Removal of Oil and Hazardous Substances, 40 C.F.R. Part 300, Subpart D, the regulation implementing Section 311 and providing more detailed guidance for oil spill cleanup operations.

5. 33 U.S.C. §§ 1321(a)(8), 1321(f)–(g), as amended and added by the Clean Water Act, Pub.L. 95-217, 95th Cong., 1st Sess. (1977). Other similar statutes addressing oil spills in special circumstances have also been enacted, for example, the Trans-Alaska Pipeline Authorization Act of 1973, 43 U.S.C. §§ 1651–1655; Deepwater Port Act of 1974, 33 U.S.C. §§ 1501–1524; Outer Continental Shelf Lands Act Amendments of 1978, 43 U.S.C. §§ 1331–1374.

6. See, for example, *Matter of Steurt Transportation Co.*, 495 F.Supp. 38, 39–40 (E.D.Va. 1980); *Commonwealth of Puerto Rico v. SS Zoe Colocotroni*, 628 F.2d 652, 673 (1st Cir. 1980), *cert. den.*, 450 U.S. 912 (1981).

7. Public Law 96-510, 96th Cong., 2d Sess. (1980), as amended by the Superfund Amendments Reauthorization Act (SARA), Pub.L. 99-499, 99th Cong., 2d Sess. (1986), codified at 42 U.S.C. §§ 9601 *et seq.*

8. 42 U.S.C. § 9601(14) (definition of hazardous substances).

9. Section 102 of CERCLA, 42 U.S.C. § 9602, which allows the U.S. Environmental Protection Agency to identify and list substances that will be deemed hazardous substances for purposes of the statute, including those not already identified by the other statutes.

10. *Wilshire Westwood Assoc. v. Atlantic Richfield Corp.*, 881 F.2d 801 (9th Cir. 1989); *City of New York v. Exxon Corp.*, 766 F.Supp. 177, 186–188 (S.D.N.Y. 1991).

11. CERCLA § 107(f)(1), 42 U.S.C. § 9607(f)(1); 33 U.S.C. § 1321(f)(5).

12. CERCLA § 107(f)(2), 42 U.S.C. § 9607(f)(2).

13. CERCLA § 101(16). See, for example, *Artesian Water Co. v. New Castle Co.*, 851 F.2d 643, 650 (3d Cir. 1988) (an aquifer is a natural resource, injury to which gives rise to a CERCLA cause of action). While these and other related terms have been defined or construed in CERCLA cases, it is likely that the same meaning would be given the terms in a Section 311 action.

14. CERCLA § 101(16), 42 U.S.C. §9601(16).

15. See *Phillips Petroleum Co. v. Mississippi*, 484 U.S. 469, 483 n. 12 (1988).

16. *State of Ohio, et al. v. United States Department of Interior*, 880 F.2d 432, 461 (D.C.Cir. 1989).

17. *In Re Acushnet River and New Bedford Harbor: Proceedings re Alleged PCB Pollution*, 712 F.Supp. 1019, 1035 (D.Mass. 1989).

18. 33 U.S.C. §1321(f)(5).

19. CERCLA § 107(f)(1), 42 U.S.C. § 9607(f)(1).

20. CERCLA § 101(32), 42 U.S.C. § 9601(32).

21. 33 U.S.C. § 1321(f)(1). In CERCLA cases, the third-party defense is further limited. See pp. 57–58.

22. CERCLA § 107(f)(2), 42 U.S.C. § 9607(f)(2); CERCLA § 301(c), 42 U.S.C. § 9651(c).

23. CERCLA § 301(c), 42 U.S.C. § 9651(c).

24. 51 Fed.Reg. 27674 (August 1, 1986) (Type-B rule), and 52 Fed.Reg. 9042 (March 20, 1987) (Type-A rule).

25. 43 C.F.R. §§ 11.40–11.41 (1987).

26. 43 C.F.R. §§ 11.60–11.84 (1986).

27. *State of Ohio, et al. v. United States Department of Interior*, 880 F.2d 432 (D.C.Cir. 1989).

28. *State of Colorado, et al. v. United States Department of Interior*, 880 F.2d 481 (D.C.Cir. 1989).

29. DOI regulations, 43 C.F.R. § 11.14(v).

30. DOI regulations, 43 C.F.R. § 11.62.

31. DOI regulations, 43 C.F.R. § 11.62(f)(2)(i)–(iv).

32. DOI regulations, 43 C.F.R. § 11.62(f)(1)(ii)–(iii).

33. *In Re Acushnet River and New Bedford Harbor: Proceedings re Alleged PCB Pollution*, 716 F.Supp 676, 686 n. 15 (D.Mass. 1984).

34. *Ohio v. DOI*, 880 F.2d at 462–463.

35. See *Utah v. Kennecott Corp.*, 801 F.Supp. 553, 566–567, 571 (D.Utah 1992) (rejecting settlement that fails to take into account existence and option values of natural resource).

36. *Ohio v. DOI*, 880 F.2d at 444.

37. 56 Fed.Reg. 19752 (April 29, 1991).

38. N.Y. Navigation Law, § 181; N.J.S.A. 58:10–23.11g.

39. DOI regulations, 43 C.F.R. § 11.14(ee).

40. N.J.S.A. 58:10–23.11f.

41. 33 U.S.C. § 1319(c)(1).

42. Migratory Bird Treaty Act, §§ 2–12, 16 U.S.C. §§ 703–711; *United States v. Engler*, 806 F.2d 425, 431–436 (3d Cir. 1986), *cert. denied*, 481 U.S. 1019 (1987).

43. 33 U.S.C. §§ 1311(a), 1319(c)(1)(A), provisions of the Clean Water Act.

44. *United States, et al. v. Exxon Corp.*, 91 Civ. 1003 (Korman) (E.D.N.Y.).

45. Correspondence attached as exhibit 2 to governments' motion for entry of settlement, dated May 24, 1991, filed in *U.S., et al. v. Exxon Corp.*, 91 Civ. 1003 (E.D.N.Y.).

46. Exhibit 3 to governments' motion for entry of settlement, dated May 24, 1991, filed in *U.S., et al. v. Exxon Corp.*, 91 Civ. 1003 (E.D.N.Y.).

47. Correspondence attached as exhibit 2 to governments' motion for entry of settlement, dated May 24, 1991, filed in *U.S., et al. v. Exxon Corp.*, 91 Civ. 1003 (E.D.N.Y.).

48. Exhibit 3 to governments' motion for entry of settlement, dated May 24, 1991, filed in *U.S., et al. v. Exxon Corp.*, 91 Civ. 1003 (E.D.N.Y.).

49. Pub.L. 101-380, 101st Cong., 2d Sess. (1990), codified at 33 U.S.C. §§ 2701–2761 (1990).

50. OPA §§ 1001–1019, 33 U.S.C. §§ 2701–2719.

51. OPA § 1003(a), 33 U.S.C. § 2703(a).

52. OPA §§ 1003(a)(3)(A)–(B), 33 U.S.C. §§ 2703(a)(3)(A)–(B).

53. H.R. Conference Rep. No. 101–653, 101st Cong., 2d Sess. 102–103 (1990).

54. CERCLA § 107(a)(4), 42 U.S.C. § 9607(a)(4).

55. OPA §§ 1002(b)(2)(A), 1006(c), 33 U.S.C. §§ 2702(b)(2)(A), 2706(c).

56. See, for example, OPA §§ 1001(20), 1006(a), 33 U.S.C. §§ 2701(20), 2706(a).

57. OPA §§ 1002(b)(2)(D), (F), 33 U.S.C. §§ 2702(b)(2)(D), (F).

58. OPA § 1002(b)(2)(C), 33 U.S.C. § 2702(b)(2)(C).

59. OPA §§ 1002(b)(2)(B), (E), 33 U.S.C. §§ 2702(b)(2)(B), (E).

60. OPA §§ 1004(a), (c), 33 U.S.C. §§ 2704(a), (c).

61. OPA § 1006(c), 33 U.S.C. § 2706(c).

62. OPA § 1006(d)(1), 33 U.S.C. § 2706(d)(1).

63. H.R. Conference Rep. No. 101-653, 101st Cong., 2d Sess. 108–109 (1990).

64. OPA § 1006(e)(2), 33 U.S.C. § 2706(e)(2).

65. 55 Fed.Reg. 53478 (December 28, 1990); 56 Fed.Reg. 8307 (February 28, 1991); 57 Fed.Reg. 8964 (March 13, 1992); 57 Fed.Reg. 14524 (April 21, 1992); 57 Fed.Reg. 23067 (June 1, 1992); 57 Fed.Reg. 44347 (September 25, 1992); 57 Fed.Reg. 56292 (November 27, 1992) and 58 Fed.Reg. 4601 (January 15, 1993). A major issue that has contributed to the delay in the rulemaking concerns the methodologies, particularly the contingent valuation (CV) methodology, to calculate nonuse values of a natural resource. Economists performing CV studies attempt to determine the dollar value of nonuse values through carefully de-

signed and administered sample surveys. Industry groups have severely criticized the use of CV methodology in natural-resource damage assessments. NOAA assembled a panel of notable economists, who concluded that CV studies could provide reliable and useful information when calculating damages resulting from nonuse (or passive-use) loss if the studies are properly designed and implemented. 58 Fed.Reg. at 4602–4614.

66. OPA § 1006(e)(1), 33 U.S.C. § 2706(e)(1).

67. 59 Fed. Reg. 1062 (January 7, 1994).

4. The Role of Conservation Organizations

Carolyn Summers

Ironically, the Exxon oil spill drew national attention to a major urban wildlife resource that few New York City residents knew existed. But for the awareness and diligence of the local New York City chapter of the National Audubon Society, one of the heron rookeries damaged by the oil spill, Shooters Island, would have been blown out of the water by the Army Corps of Engineers some 10 years previously. After defeating this initial threat, the New York City Audubon Society (NYCAS) created the Harbor Herons Project to protect and manage these regionally significant rookery islands as a wildlife sanctuary complex (TPL/NYCAS 1990). Working with the cumbersome New York City bureaucracy, in 1985 NYCAS arranged for the transfer of Pralls Island (the rookery most heavily damaged during the Exxon oil spill) to the New York City Department of Parks and Recreation (NYC Parks) and for its designation as a wildlife sanctuary.

In addition, NYCAS signed a management agreement with NYC Parks to manage Pralls Island for its continued use as reproductive habitat by the colonial wading birds. To provide a sound scientific basis for management decisions, in 1985 NYCAS entered into an agreement with Manomet Bird Observatory (Manomet) to monitor the reproductive success of the wading-bird colonies for a period of 30 years. This farsighted management decision was to provide, ultimately, some of the best available baseline documentation of the significance of the wildlife resources imperiled by the Exxon oil spill.

The agreement by NYCAS to manage Pralls Island fed the chapter's growing awareness of the need for comprehensive protection programs for the key remaining natural areas within New York City. Under the leadership of then–conservation chair Albert F. Appleton, NYCAS formed a partnership with the Trust for Public Land (TPL), a national nonprofit conservation organization dedicated to preserving open space, to produce a comprehensive inventory of lands needed to buffer the tremendous wildlife and other natural resources found in New York City's Jamaica Bay. The resulting report, *Buffer the Bay: A Survey of Jamaica Bay's Unprotected Open Shoreline and Uplands* (TPL/NYCAS 1987), established the credentials of both the NYCAS Urban Wilds Program and TPL's New York City Land Project in local land use planning and preservation and has become a model for subsequent land-planning and preservation studies.

The success of *Buffer the Bay* led NYCAS to further expand the scope of the Harbor Herons Project. In 1989, NYCAS contacted TPL to initiate another cooperative venture modeled on *Buffer the Bay*, this time to highlight the unique natural resources located on Staten Island, resources that support the colonial wading-bird colonies. As with the Manomet data on population dynamics of the herons, the decision to inventory and document the resource base for Staten Island's portion of the Arthur Kill ecosystem was ultimately to play a critical role in the litigation and subsequent settlement negotiations that resulted from the Exxon spill.

TPL began a comprehensive inventory of the land-based habitat utilized by the colonial wading birds, with a view to establishing the need for additional protected areas. The TPL/NYCAS Harbor Herons study examined existing land uses and property ownership patterns, analyzed projected land use trends and threats to existing open space, and inventoried natural resources with an emphasis on the use of those resources by the birds. The boundaries of the study area were dictated by existing patterns of urbanization of Staten Island, political boundaries, and the location of wetlands and rookeries rather than by the potential maximum flight ranges of the waders. For instance, although excellent foraging habitat existed across the Arthur Kill in New Jersey, as well as on Staten Island, and although a future New Jersey study was contemplated, for pragmatic reasons the decision was made to limit the scope of the initial study to Staten Island.

The Spill: First Responses

The Arthur Kill oil spill occurred during the final stages of the Harbor Herons study; in fact, the report of the study's results had already been through a number of edits. Because of the working relationship between TPL staff and the previous years' field biologists from Manomet and Rutgers, the spill was reported to TPL on the morning of January 2, 1990, by Rutgers biologist John Brzorad, who was on the scene to witness the initial devastation firsthand.

Through this information and subsequent firsthand reports, staff at TPL were able to gauge the accuracy of early reports by the local media. Newspapers, television, and radio were just beginning to pick up on what was to become a major news event. TPL staff were outraged by these early reports because there already appeared to be an effort at "spin control," downplaying the amount and potential effects of the oil. More disturbing was the consistent theme that if a spill had to occur in the New York–New Jersey harbor, the Arthur Kill was probably the best place due to its severely degraded nature. No concern for the vulnerable tidal wetlands, the Pralls Island rookery just across the Arthur Kill from Exxon's Bayway facility, or other wildlife was voiced in any of the early reports.

By late afternoon of January 2, it was already clear to TPL staff that the record needed to be set straight. Calls were placed to all of the local media, alerting them, first, to the existence of the harbor herons and, second, to the serious threat this oil spill posed to the continued viability of the habitat that supports this large breeding population. Information provided by TPL was the first to contradict the official reports that the oil spill had occurred in a "dead sea." As the story heated up, the media became anxious to exploit this new angle, that far from occurring in a wasteland, the Exxon oil spill had in fact taken place right next to a regionally significant wildlife sanctuary.

During the first week of the spill, several circumstances combined to reinforce TPL's new role as principal media contact for the Harbor Herons Project. First, project staff were able to cite the extensive research that went into both TPL's and Manomet's harbor herons studies (see Chapter 13). Second, because of the working relationship among the two other most-interested parties, NYCAS and Manomet, TPL staff were able to effectively direct reporters to those local experts best able to field particular questions. Third, TPL maintained a fully staffed office with all of the basic requirements for a media campaign: multiple phone lines, receptionist, fax machine, and press representative. Other local conservation groups were virtually all volunteer, and many of these volunteers held demanding daytime jobs. Thus, the response largely fell to TPL. Even with a paid, full-time staff and basic equipment, TPL found its resources strained to the limit. The sheer volume of phone calls threatened to overwhelm both the receptionist and the small cadre of staff able to respond to the calls.

During the process of educating the media, the TPL/NYCAS/Manomet Harbor Herons team received its own education in media crisis management. Misquotes were rampant and caused a great deal of anxiety and confusion for participants. The groups were frustrated that although overall attention was soon shifted from the original "spin control" version to the actual effects of the oil on certain of the Arthur Kill's vulnerable resources, some of the more mainstream media, most notably the *New York Times*, appeared almost to ignore the entire event. Even with some successful coverage of the plight of the oiled birds and the threat to the returning herons, the other valuable resources of the Arthur Kill were ignored by the media.

One of the most important outcomes of TPL's media outreach was the subsequent response to TPL from other conservation groups, local and national. The resulting network of conservation groups strengthened the media awareness campaign. Effective media contributions, particularly from the American Littoral Society (ALS) and the Natural Resources Defense Council (NRDC), complemented and expanded upon TPL's original media message.

Networking extended into the field so that not a day went by without firsthand accounts of the oil spill and its impacts. These accounts were relayed among the conservation groups so that no matter which group was contacted by the media, reports were as accurate as possible. When time permitted, press releases were frequently shared among the groups prior to release. This ongoing cooperation laid the foundation for a successful coalition among the most active organizations, now known as the Safe Harbor Coalition, which continues to monitor the disposition of the Exxon settlement funds.

Since most of the official early reports emanated from the Coast Guard, TPL staff contacted them on the first day of the spill to provide information regarding the resources most in need of protection and to learn the Coast Guard's view of the progress of the spill. Once the phone numbers of the "central command" had been tracked down, TPL obtained regular updates on the progress of the spill and containment efforts. In addiition, TPL staff were anxious to make the Coast Guard, which was in charge of the containment effort, aware of the vulnerability and significance of the resources impacted by the oil spill. Although the Coast Guard spill response maps identified the rookery islands and adjacent tidal wetlands as ecologically sensitive areas, those in charge did not appear to be aware of, or were not acting on, this information.

The Coast Guard was initially less receptive than the media had been to TPL and NYCAS's insistence that these areas are ecologically sensitive and require special treatment and handling. This attitude may have been partly due to the fact that certain response decisions made by the Coast Guard early on had aggravated the effects of the spill on Staten Island's tidal wetlands. Fortunately, the local government agencies worked with the conservation organizations to impress upon the Coast Guard the need to protect the tidal wetlands.

At the beginning of the spill, the conservation groups widely perceived that the Coast Guard was taking most of its cues from Exxon, once Exxon had finally accepted its responsibility. The conservation groups began to demand more input into containment decision, as did local government agencies. Because of the overlap in jurisdictions (two states, two municipalities, and three federal agencies), confusion reigned throughout the first critical days as the agencies sorted out their respective roles.

Although the environmental groups depended largely on local governments to direct the Coast Guard to protect the wetlands, they did not neglect personal pressure through daily phone calls to the spill command center. Thanks to the team of environmentalists in the field, TPL was able to relay to the command center containment recommendations based on direct observations. Although the government agencies collectively staffed spill response around the clock, these were not large staffs to begin with, and the magnitude of the spill and the area it covered nearly overwhelmed the available resources of the various agencies.

Another significant outcome of the environmental media campaign was the outpouring from all over the region of volunteers who wanted to become involved in the wildlife rescue efforts. Begun almost immediately by the Manomet and Rutgers field biologists as well as other local scientists, residents, and environmental activists, the initial rescue attempts were greatly hampered by the inaccessibility of the area, the lack of boats and other rescue equipment, and, of course, the spill itself.

Within the next few days, the governments and Exxon mobilized additional resources to recover wildlife. Fumes from the no. 2 fuel oil permeated the entire area, making exposure for extended periods hazardous. Many dedicated individuals involved in the spill response and wildlife rescue became quite ill, with symptoms ranging from bronchial infections resulting in complete loss of voice to severe headaches and dizzy spells that in some cases continued long after the initial exposure.

Although the volume of calls from wildlife rescue volunteers was an additional strain on TPL's limited resources, it was heartening to know that so many people were horrified by the spill and felt the need to make a meaningful contribution to the response. Many members of NYCAS also called their local office to volunteer. Their names and phone numbers were referred to TPL staff to maintain coordination of efforts. The names and phone numbers of those volunteers able to work during the first week were given to the Manomet and Rutgers field biologists who daily manned NYCAS's boat—the only boat available to the conservation groups. Since most volunteers were able to participate only on the weekend, the sporadic attempts at rescue during the first week were followed by a massive attempt, jointly organized by TPL and NYCAS, to survey the entire area on January 7, the first Saturday after the spill.

The day after the spill, Exxon brought in Tri-State Bird Rescue & Research, Inc., a recognized wildlife rehabilitation organization, to deal humanely with, and when possible rehabilitate, the increasing numbers of distressed and dying wildlife that were being found. Unfortunately, Exxon established the rescue center at its Bayway facility on the New Jersey side of the Arthur Kill, a forbidding and inaccessible location. Not only was the entrance difficult to find, but excessive security made entry, even with distressed wildlife, difficult at best. The operation was set up so as to be easily accessible by water, since the majority of Exxon's activities were waterborne. This was no help to residents of the surrounding communities, most of whom had no access to a boat. The location of the rescue center necessitated extensive coordination with Exxon to ensure the success of the January 7 volunteer rescue effort.

In addition to working out a system with Exxon for the retrieval of wildlife and their conveyance to Bayway, a great deal of logistical preparation for the rescue effort was necessary. Lack of boats was the most severe problem, but a local Staten Island environmentalist provided two row-boats from the Boy Scouts, and one couple brought their own canoe. Boats were necessary because only certain areas could be walked, as the foot traffic would have compacted and further damaged the delicate structure of the peat underlying the already-oil-soaked tidal marshes. During the first week, rescue efforts largely concentrated on the shorelines of the Arthur Kill and Pralls Island, and it was decided that the three major tidal creeks within a mile of the spill should be targeted. As it turned out, there were just enough boats.

Arrangements were made for a location where teams would be formed, maps and instructions given out, and supplies distributed. Detailed directions to this isolated meeting place were needed to enable the volunteers, media participants, and government leaders to find it. Instructions on what to wear and what to bring were also given out in advance. Some efforts was made to discourage people who might not have been in optimum health from participating by informing them of the difficulties of the terrain and the hazards from the oil fumes, which still pervaded the atmosphere.

From the outset it was made clear to all participants that the rescue mission had a dual purpose: to rescue wildlife and to collect evidence of damage to wildlife and the environment. Volunteers were encouraged to bring glass bottles to collect water samples and notebooks to report the day's events. Each team appointed one person to take notes, although individuals often made their own additional notes. Carcasses of dead wildlife were to be picked up, not just for the sake of evidence, but also to prevent scavenging by gulls and hawks. All wildlife, whether dead or alive, had to be turned over to Exxon employees at a designated staging area.

Once final instructions, maps, and supplies were given out, all participants drove to the staging area, located in an even more isolated area, at
the back of a private industrial facility. After everyone was shown this
location, the teams dispersed to their appointed geographic area. One
boat team was allocated to each of the three major tidal creeks near the
site of the spill. Another team was assigned to an Exxon boat and crew
and partially covered ferrying the wildlife across the Arthur Kill to
Bayway. Some volunteers took the NYCAS boat with the Manomet and
Rutgers field biologists and covered Prall's Island. Two additional teams
walked areas along the tidal creeks where berms or other fill material
made a firm pathway.

The walking teams were at the greatest disadvantage in the rescue
effort. Although all teams were hampered by lack of equipment such as
nets, this was more of a problem for teams without boats, as it was nearly
impossible to follow ducks that took to the water. One team experienced
the anguish of carefully stalking an oiled duck along the shoreline, never
able to come quite close enough to grab it, until it finally jumped into
the water and quickly sank, never to reappear.

Volunteers who were successful in capturing injured wildlife often
faced a long, frustrating wait at the staging area. Since the volunteers
were not provided with two-way radios, Exxon boats were supposed to
periodically stop by the staging area to make pickups. This arrangement
was not the most efficient use of the volunteers' time and added to the
distress of the wildlife, which ideally should have been turned over to
Tri-State as quickly as possible. In addition, some of the volunteers
expressed misgivings about turning over dead ducks and other wildlife
to Exxon personnel. All volunteers kept their own list of what they had
found, but since Exxon was merely keeping a running tally by species,
there was no way to be sure that individual carcasses turned over to
Exxon would wind up on the official list.

All in all, despite the adversities encountered, the rescue effort was a
success in terms of wildlife rescue and damage assessment, and in terms
of media coverage—the event was covered by television and newspapers. Several distressed birds were taken in to Tri-State. Many more
were, unfortunately, discovered oiled and dead, but at least their bodies
did not serve as a meal for some unfortunate predator, serving instead as
evidence of the damage inflicted by the oil spill. The field notes taken by
the volunteers proved to be excellent. Detailed and accurate, and in
some cases accompanied by hand-drawn maps indicating locations of
samples taken, wildlife found, and, most important, extent of oil. These
field notes were ultimately used by the governments' damage assessment team and, together with the field notes of the New York Depart-

ment of Environmental Conservation's (DEC) personnel, provided the best documentation of both the geographic extent and severity of oiling on the Staten Island side of the Arthur Kill.

Cleanup and Damage Assessment

On January 18, approximately 2 weeks after the spill, a joint meeting was scheduled among the conservation organizations and the governments to discuss cleanup options, protection of resources during cleanup operations, and the damage assessment. The meeting was very well attended by scientists and lawyers from three federal agencies, the two states, and New York City. The conservation organizations had contacted as many oil spill experts as possible from all over the country to obtain the latest information on cleanup technologies and damage assessment procedures. This information, together with the detailed resource information collected as part of TPL's Harbor Herons study, was discussed among the participants. This informal exchange of information marked the beginning of the Exxon Oil Spill Advisory Committee, a coalition of conservation organizations that later became known as the Safe Harbor Coalition, which thereafter scheduled regular meetings with the governments to provide input into cleanup and other decisions.

With formation of the advisory committee spreading the work among the various conservation organizations, TPL determined that its most important task was to finish and release its Harbor Herons study. TPL felt that the report would aid the damage assessment, stimulate additional public awareness of and support for the local resources, and set the agenda for a restoration/acquisition program financed by the proceeds of a lawsuit against Exxon. An ambitious review, editing, and production schedule was circulated to the staff of the New York City Land Project with the goal of submitting the manuscript to the printer by the end of February.

By sticking to the demanding schedule, TPL was able to release *The Harbor Herons Report: A Strategy for Preserving a Unique Wildlife Habitat and Wetland Resource in Northwestern Staten Island* (TPL/NYCAS 1990) on March 12, 1990. Due to the notoriety of the Exxon oil spill, the report received more than the usual attention accorded the release of a land use planning report. The report was circulated widely and was well received by local government planners, as well as by those involved with the Arthur Kill oil spill. The proof of its usefulness as a tool was the frequency with which government representatives, conservation organizations, and even Exxon representatives referred to it at oil spill meetings.

One of the most important recommendations made by the report was that the Arthur Kill resource base be viewed as one integrated ecosystem. Given the baseline information available for the harbor herons, it was logical that the birds be the focus of the damage assessment; however, in meetings with the governments, the advisory committee consistently raised the point that the entire ecosystem supports the breeding population. In addition, the fish, invertebrates, and reptiles are important resources in their own right and need to be taken into account during the damage assessment. The advisory committee maintained that all of the biological resources of the Arthur Kill are unique and contribute to local biodiversity, strengthening and stabilizing the entire ecosystem.

Although the governments were initially responsive to the recommendations of the advisory committee, follow-up was complicated by the sheer number of governments involved. Concerns for local industry still seemed to dominate decision making, for example, the Arthur Kill was prematurely reopened to ship traffic, washing more oil onto Pralls Island and the marshes. The cleanup remained at a very conservative level in terms of risk and technologies used. The conservative approach, determined by the Coast Guard, dictated that the marshes not be physically cleaned by any agent other than natural tidal ebb and flow, because human traffic was thought to cause more damage than the remaining oil.

Unfortunately, the time lag for decisions did not benefit the damage assessment. By February, no official damage assessment was in progress, partially because of uncertainty regarding how to fund it. Acting independently, local scientists from Rutgers University and Ramapo College initiated studies of the effects of the oil on fish and invertebrates within days of the spill. Although never fully funded, these efforts resulted in published baseline studies (see Burger, Brzorad, and Gochfeld 1991, 1992; Chapter 9).

Progress toward an official damage assessment was made when New Jersey sued Exxon; under New Jersey's spill bill, Exxon was required to pay for the governments' damage assessment. After the funding was cleared up, questions regarding which natural resources to include in the damage assessment and which special consultants were needed slowed the process of seeking out a consultant team to handle the damage assessment. Chemical and other evidence that could have been gathered within the first 2 months before the oil had weathered considerably was lost. Sediment sampling was not actually begun until late March.

The advisory committee expressed concern that the damage assessment, once underway, excluded certain resources, particularly fish, reptiles, and the salt marsh cordgrass *Spartina alterniflora*. The governments, however, partially on the advice of the special economic consultant, based the damage assessment on the food chain supporting the harbor herons. This meant that the only fish and invertebrates to be assessed were species preyed upon by the wading birds, and the condition of the marshes was not analyzed at all.

One of the major tasks of the damage assessment team was to produce a definitive map of the extent of the Exxon oil spill. The first draft produced by the damage assessment consultants was woefully inadequate on the Staten Island side of the Kill. The first week or so after the spill, New Jersey Department of Environmental Protection (NJDEP) staff walked the entire New Jersey side of the Arthur Kill and made an official map of the extent of the oil spill. This was never done on the New York side. The consultants worked from the Coast Guard maps along with the NJDEP map, with the result that damage to all of the tidal creeks on the Staten Island side was omitted. The only documentation of the thorough contamination of the creeks on Staten Island was field notes, most of which came from the conservation group volunteers. The information from all field notes was pooled by government representatives and the damage assessment consultants to complete the picture of the areal extent of the spill. This vital information added to the damage assessment not only additional acreage covered by oil but also the most sensitive type of acreage—salt marsh (Fig. 4.1).

The Settlement Negotiations

Even before the damage assessment results were in, lawyers from the five governments that had filed lawsuits against Exxon (the federal government, states of New Jersey and New York, and cities of Elizabeth and New York) and Exxon's lawyers had begun to negotiate a settlement. For valid reasons, the governments were not able to discuss the details of the negotiations with the advisory committee. This proved to be a frustrating period for the members of the committee and others interested in conservation. Unfortunately, the degree of trust in the governments' legal team was low, not because of the level of competence of the government attorneys, but rather because of the perceived attitude of disdain for the resources of the Arthur Kill displayed by a few key members of the legal team. The committee feared that such an attitude would ensure a low settlement.

A notable exception to this attitude was shown by the New York City Counsel, who displayed a solid commitment to the resources and who understood the need to work with the advisory committee. Although the

Figure 4.1. Healthy *Spartina alterniflora* along Yellow Craven Creek before the spill, in the summer of 1989 (top), and after the spill, in the summer of 1990 (bottom). Note stubble of dead *Spartina* along eroded banks.

Federal Department of Justice was designated to contact the advisory committee, New York City's lawyers often conveyed such information as could be released to the advisory committee chair, who was at NRDC. Working with whatever information became available, the advisory committee continued to lobby the governments for a strong settlement that would set the tone for future spills. Memos were written from the advisory committee to the governments detailing items that the committee felt should be included in the final settlement. At one stage in the negotiations, the committee chair was requested by the New York City Counsel, on very short notice, to prepare a list of spill prevention measures that the committee would like to see Exxon undertake as part of the settlement package.

In spite of difficulties, the advisory committee was able to influence the course of events at many key points by contacting individual government officials during the negotiation process. Ultimately, a successful settlement was reached. (For complete details of the settlement, see Chapter 3.) Although some felt that a judge and jury might have awarded more money, on balance, all participants agreed that it was better for the sake of the resources to have money to be able to restore and acquire land now instead of years later. For the authors of *The Harbor Herons Report*, the settlement fulfilled their highest hopes. Many aspects of the report were incorporated into the settlement documents, and the report continues to set the agenda for the expenditure of Exxon funds within the Harbor Herons Wildlife Refuge. Land is being purchased that will preserve habitat for the diverse flora and fauna that exist within the Arthur Kill ecosystem. NYC Parks has been funded to manage and restore the damaged marshes in newly acquired land on Staten Island. The advisory committee, now the Safe Harbor Coalition, continues to monitor the expenditure of Exxon funds for land acquisition and restoration on both sides of the Arthur Kill.

Preparing Conservation Organizations for Future Oil Spills

As this country is dependent on petroleum products, it is clear that any community can experience an oil spill (NRDC 1990). Local conservation organizations can take steps to prepare for such an event, and may, in some instances, be best able to take charge in an emergency (Table 4.1). The government agencies charged with spill control are, in general, underfunded and understaffed, and they must work through a bureaucracy. Conservation organizations can act quickly. In a major spill such as that in the Arthur Kill, all participants can provide useful services.

The Oil Pollution Act of 1990 provides for the establishment of area committees to be set up by the captain of each port zone of the United

Table 4.1. Response to Oil Spills

Timetable	Activities
Prespill	Inventory and document natural resources.
	Obtain information on effects of oil, cleanup techniques, methods for collection of evidence, and government response.
	Attend wildlife rehabilitation workshops, devise protocol for volunteer wildlife rescue.
	Participate as ex officio members of Coast Guard local-area committees.
Spill	Inform governments about affected or threatened natural resources.
	Coordinate volunteer response efforts with appropriate government agencies.
	Raise public awareness regarding the spill and its effects on local natural resources.
	Observe protocols for field notes, collection of evidence, wildlife rescue, and protection of resources.
	Monitor containment and cleanup activities.
Postspill	Share collected information with government agencies.
	Assist damage assessment process.
	Monitor cleanup and restoration activities.

States. These committees are charged with drawing up a comprehensive spill response plan for that area. Although membership on the committees is restricted to government agencies, ex officio membership by conservation groups is supported by the Coast Guard.

Participation by conservation groups in the work of the committees is vital. The area plans should describe the roles and responsibilities of the various parties during response to an oil spill. The role played by volunteers must be developed within the framework of the plan so that coordinated, effective use of volunteers will be ensured. In addition, the area plans should describe environmentally sensitive areas that must be protected and so-called sacrifice areas where oil can be corralled to prevent spreading. Facility and vessel response plans will depend on these descriptions in planning for spill responses, as will government agencies and the on-scene coordinator. It is critical for conservation organizations to become involved with the preparation of the area plans by joining as ex officio members of an area committee.

As the Arthur Kill experience proves, perhaps the single most important contribution that conservation organizations can make is to have on hand detailed knowledge regarding local natural resources. Not only will this make possible immediate decisions regarding containment and cleanup options, but baseline information on wildlife and other natural resources is vital to a complete damage assessment after a spill. Thorough knowledge of the local resource base and the ability to communicate that knowledge effectively is the best preparation for an oil spill.

Although a comprehensive report such as *The Harbor Herons Report* becomes a valuable tool in an oil spill, particularly the later stages, organizations should not be put off if they lack descriptions of resources before a spill. Sources of information are available to conservation organizations. *The Harbor Herons Report* itself relied heavily on previous natural-resource work by government agencies, notably the Department of Environmental Conservation's wetlands inventory, local institutions, and private individuals. Often more information exists than expected and needs mainly to be organized in a useful fashion.

The most useful organization of natural-resource information for an event such as an oil spill is graphic, preferably a series of large-scale display maps. These should provide as many overlays of resource information as possible without becoming so cluttered as to be unreadable. The backup documentation for such maps can be filed in any number of ways as long as the information remains accessible. The importance of mapping the local resource base cannot be overemphasized. Such a map will, in an emergency, become a handy tool for geographic decisions that must be made as quickly as the oil itself will travel. On the other hand, no one will be able to take the time to read even the most carefully written synopsis on sensitive natural resources reportedly located somewhere in the path of destruction. Maps and other graphic displays are also useful for the media.

If no maps or other easily digestible information are readily available, government agencies will most likely take a conservation group's word for the existence of local natural resources that their own inventories may have overlooked. Be persistent. To be effective, it is important to establish credibility as an advocate early in the spill response process.

Oil spill response and cleanup activities have evolved a wide range of possible options, some more controversial than others. To participate effectively in the decision-making process, it is desirable to have in advance a working knowledge of oil spill response products and procedures. Conservation organizations should be able to obtain this information through the government agency locally responsible for emergency response. Information on this topic can also be obtained from national environmental organizations working on oil issues, such as the NRDC, as well as from the International Tanker Owners Pollution Federation or the U.S. Coast Guard National Response Center.

Adding conservation volunteers to the spill response effort in the field may be appropriate but should be coordinated through the official response agencies, for the health of the environment as well as of the individual volunteers. The more knowledge the organization has accumulated with regard to certain emergency procedures and protocols, the more effective its efforts will be. Before participating in the field, volunteers may need to be informed about specific resource protection needs, for example, that walking single file in a marsh is necessary to prevent additional damage through compaction. In some cases access must be even more severely restricted. Conservation organizations must be aware of the protection from human disturbance needed by local resources and the potential damage that can be caused by response activities.

Wildlife rescue is another area in which advance preparation and education can make all the difference. If the oil spill is near an established wildlife rehabilitation center, the job is greatly simplified. Most communities are not so lucky. However, many conservation organizations know of one or two local individuals who are licensed wildlife rehabilitators. Local zoos, veterinarians, and humane societies may also be willing to open their doors in an emergency; however, it is better to find out their policy in advance. Determine what it would take to enable local rehabilitators to adequately respond in an emergency: the type of facilities and equipment that would be needed, number of volunteers, and so on. Visit an established facility to get an idea of what is involved. To have enough trained volunteers, it may be necessary to sponsor a seminar on wildlife rehabilitation techniques. A less technical seminar on first aid in the field for rescue volunteers is also useful. Homework of this sort may enable your organization to retain control of the wildlife rescue effort and to avoid the appearance, such as occurred in the Arthur Kill spill, of a conflict of interest, in that Exxon was allowed to exercise almost complete control over the wildlife rescue effort.

Remember that any fieldwork during an oil spill is a two-pronged effort. A conservation organization may feel that its first priority is to alleviate the suffering of distressed wildlife, but individual volunteers must not lose sight of the valuable opportunity to gather concrete evidence that will form the basis of a future damage assessment. Although picking up wildlife corpses is an ordeal, doing so prevents further progress of the oil through the food chain as well as provides a form of evidence. As simplistic as it may sound, every damage assessment begins with a body count. And for every wildlife corpse recovered, hundreds more are presumed to have died unrecovered.

Do not concentrate exclusively on the more "charismatic macrofauna," such as ducks and otters, when gathering evidence. Invertebrates, fish, reptiles, and amphibians are all important in their own right as well as within the food chain and are subject to the same type of body count analysis in a damage assessment. It is particularly important to document rare and endangered species, as they "score" higher in any damage assessment.

Official protocols exist for dealing with any evidence of damage. "Chain of custody" refers to the establishment of a paper trail covering the transfer of physical evidence such as the corpse of an oiled duck from the original field personnel through other individual handlers to its ultimate destination, usually a government wildlife pathologist. The wildlife pathologist will be able to determine whether or not oil contamination was the cause of death. This may seem obvious when dealing with a carcass coated with and reeking of oil; however, to withstand legal scrutiny positive proof is required.

At the same time volunteers are in the field rescuing wildlife, they can pick up samples of oil, either in sediment or in water, that will later be matched with the official oil sample and used to definitively establish the geographic extent of oil coverage and the severity of contamination at that specific location. It is particularly important to document the spill extent. All volunteers should make notes, with exact dates, of any evidence of oiling, and these notes should result in a map of the outline of the geographic extent of the spill. If possible, locations of samples or other evidence taken should also be indicated on the spill extent map.

Sampling protocols differ slightly according to type of oil and other factors, but the use of glass bottles is always safe. Plastic containers can leach contaminants into an oil sample. All containers should be labeled, using a crayon or other waterproof implement, with the approximate location in the field, date, and name of collector. Oil samples should be frozen as soon as possible, as the breakdown of certain types of oil, such as no. 2 oil, into their chemical constituents can be quite rapid.

More detailed information on the official protocols referred to above can be obtained in advance of an oil spill by contacting the regional offices of the U.S. Fish and Wildlife Service and the Coast Guard. Do not be put off, however, by a lack of such information in the event of an emergency. Detailed field notes can provide adequate evidence and, as noted, played a vital role in the establishment of the spill extent for the damage assessment of the Arthur Kill oil spill.

Conclusions

Early participation by local conservation groups was critical to all phases of the Arthur Kill oil spill. Through the media, conservation groups raised public awareness regarding the special nature of the wildlife resources of the Arthur Kill ecosystem. Continued media contacts focused on the long-term threat to a valuable habitat from the oil spill. The consequent media pressure on Exxon ensured the expenditure of greater efforts by the company to contain and clean up the spill in an effort to counteract negative publicity. Similarly, media pressure influenced the efforts of the lead government agency responding to the spill, resulting in a shifting of emphasis from attention to local industry to a more protective approach toward the natural resources.

Working with local scientists, conservation organizations provided firsthand knowledge of the Arthur Kill's natural resources. This knowledge formed the basis of the damage assessment. Advocacy by the conservation groups for the entire Arthur Kill ecosystem, buttressed by solid research and documentation, encouraged the governments' litigation team to fight Exxon's well-funded but ultimately unsuccessful efforts to downgrade the value of the Arthur Kill's resources.

Acknowledgments

The author dedicates this chapter to the committed conservation volunteers who form the backbone of environmental protection and preservation efforts in this country and particularly wishes to acknowledge the Arthur Kill wildlife rescue volunteers. In addition, this chapter could not have been written without the support of fellow team members Mary Kearns-Kaplan, John Brzorad, and Alan Maccarone.

The author especially wishes to thank Nina Sankovitch, chair of the Safe Harbor Coalition, and Albert F. Appleton, commissioner of the New York City Department of Environmental Protection, for expert comments on this manuscript. In addition, the author wishes to recognize Al Appleton for his vision, leadership, and strength throughout all phases, past, present, and future, of the Harbor Herons Project.

Literature Cited

Burger, J., J. Brzorad, and M. Gochfeld. 1991. Immediate effects of an oil spill on behavior of fiddler crabs (*Uca pugnax*). *Arch. Environ. Cont. Tox.* 20:404–409.
———. 1992. Effects of an oil spill on emergence and mortality in fiddler crabs (*Uca pugnax*). *Environ. Monit. Assess.* 22:107–115.
Natural Resources Defense Council. 1990. *No Safe Harbor: Tanker Safety in America's Ports.* New York: Natural Resources Defense Council.
Trust for Public Land/New York City Audubon Society (TPL/NYCAS). 1987.

Buffer the Bay: A Survey of Jamaica Bay's Unprotected Open Shoreline and Uplands. New York: Trust for Public Land in conjunction with New York City Audubon Society.

————. 1990. *The Harbor Herons Report: A Strategy for Preserving a Unique Wildlife Habitat and Wetland Resource in Northwestern Staten Island.* New York: Trust for Public Land in conjunction with New York City Audubon Society.

5. Rehabilitation of Contaminated Wildlife

Lynne Frink

On January 2, 1990, Tri-State Bird Rescue & Research, Inc., was contacted by Exxon, under directive from the U.S. Fish and Wildlife Service, to manage a response for wildlife contaminated by the previous night's oil spill in the Arthur Kill. Since the oil spill occurred in winter, the primary bird species at risk included wintering waterfowl and resident gull species. A complete list of oiled birds that were retrieved dead following the oil spill can be found in Table 13.1.

Tri-State Bird Rescue and Research is a federally licensed, professionally staffed wildlife rehabilitation and research center located in Newark, Delaware. Tri-State was founded in 1976 to study the effects of oil on wildlife, design effective response mechanisms for wildlife affected by oil, and determine treatment protocols. Although Tri-State's scope has broadened beyond oil-contaminated wildlife, it has gained an international reputation for its oil spill response and training capabilities. Tri-State conducts training seminars and workshops throughout the United States and Canada for industry, federal and state government, and academic institutions. It has presented two international conferences on the effects of oil on wildlife. In 1991, the United Nations requested Tri-State to organize a six-person team to provide training and response assistance following the 250-million-gallon spill in the Persian Gulf.

This chapter provides a general overview of the effects of oil on birds and discusses current treatment protocols. It then presents an overview of the wildlife rehabilitation effort following the 1990 Exxon spill in the Arthur Kill and discusses the special problems associated with this effort.

Wildlife rehabilitation focuses primarily on the adverse physiological effects of oil on individuals. These effects, while complex, can often be successfully counteracted through the cooperative efforts of veterinarians, biologists, and rehabilitators with oil spill response experience, though often determining and carrying out treatment protocols require experience and technical skills not within the general expertise of veterinarians and biologists. The primary objective of wildlife rehabilitation is to care for injured animals and release them to their natural environment. Wildlife rehabilitation fills two purposes in oil spill response: philosophically or morally, it provides a humane response to wild animals harmed through human activities; biologically, it attempts to treat and return injured animals to healthy breeding populations in the wild.

Rehabilitation efforts can be particularly important when endangered or threatened species are involved.

The Effects of Oil on Birds

The general effects of oil on birds can be characterized as environmental, external, and internal.

Environmental Effects

This is perhaps the broadest category of the effects of oil on wildlife. Environmental effects include, but are not limited to, immediate contamination of the food source biomass, reduction in the animals and plants that form the food sources, contamination of nesting habitat, and reduction in reproductive success through contamination and reduced hatchability of eggs or temporary inhibition of ovarian function (Albers 1977, 1991; Leighton 1991).

Environmental effects can be acute and temporary or chronic and long-lasting and can involve entire populations of birds. A tidal marsh can be heavily contaminated by oil early in a spill, but much of the damage can be temporary, in that certain types of oil may flush out of the marsh after one or two tidal cycles. On the other hand, more permanent damage can be incurred following oil contamination of adults and eggs at a nesting colony of birds like common murres, pelagic birds that do not breed until they are 4 years old, and lay only one egg a year. Damage to this colony could have a long-lasting effect on regional populations.

In a number of spills, the environmental effects of the oil may be minimal while the physiological effects on the individual animals in local populations may be acute and life-threatening.

External Effects

These effects are the most noticeable and most immediately debilitating (Dein and Frink 1986). Birds that are most often affected by oil spills include those that remain on the water, such as ducks, loons, and grebes; and those that feed in the water, such as gulls, terns, and herons, and birds of prey like bald eagles and ospreys. Oil can contaminate the entire bird or only parts of the bird, depending on the amount of oil in the water and the bird's natural behavior (swimming, wading, diving) in the water. By disrupting the microscopic structure of feathers, oil destroys the waterproofing and insulating properties of plumage. The oiled bird may suffer from chilling, be unable to fly, or be unable to remain afloat in the water. An oiled bird has difficulty obtaining food or escaping predators.

In addition to the decreased foraging ability of the animal, the presence of oil in the environment usually results in the loss of available food. It is the external effects of oil that are most often noticed by the general public, and treatment efforts in the past have concentrated on removing oil from feathers.

Internal Effects

The internal effects of oil on birds, while not as apparent, are equally life-threatening. Direct toxic effects on the gastrointestinal tract, pancreas, and liver have all been documented (Leighton 1991). Ingestion of oil by birds attempting to clean feathers through preening frequently results in ulceration and hemorrhaging of the lining of the gastrointestinal tract, inhibiting digestive and absorptive abilities (Langenberg and Dein 1983). The animal's damaged digestive system cannot utilize food or water. A similar irritation of mucosal surfaces can lead to ulceration of the conjuctiva and corneal surface of the eyeball, and of the moist surfaces of the mouth. Oil aspiration pneumonia is not uncommon in oiled birds and can occur when birds, attempting to clean their feathers through preening, aspirate droplets of oil. Severe and fatal kidney damage has been documented (Leighton 1991). The debilitated bird also becomes susceptible to secondary bacterial and fungal infections that are also life-threatening (White 1991).

The internal effects are not taken into account by the general public and by many people attempting to rehabilitate oiled birds.

Treatment of Oiled Birds

This section provides a general overview of the facility needs and treatment protocols for the rehabilitation of oiled birds. Specific information about the Arthur Kill spill follows this general overview.

Stabilizing the Bird

As soon as possible after retrieval, efforts are made to stabilize an oiled bird by treating the internal effects of the oil (Table 5.1). Stabilization efforts may be initiated in the field, but complete medical care is provided when the animal arrives at the clinic. In addition to a complete standard physical examination, each bird receives a temporary, numbered plastic leg band, and individual records are kept on each animal. Weight and cloacal (rectal) temperature are recorded. Oil is removed from the mouth and the nares (nostrils). The vent is checked for oil and matted feathers, which might cause cloacal impaction. The eyes are flushed with warmed, sterile saline or ophthalmic irrigation. Small blood

Table 5.1. Overview of Field Stabilization Protocols for Nonpelagic Species

Active, alert birds

1. Check for bleeding and traumatic injuries. Treat if present.

2. Take cloacal temperature (normal = 102–106° F) and weight.

3. Assess condition of plumage and skin.

4. Flush eyes with warmed sterile physiologic saline.

5. Clean mouth and nares.

6. Administer warmed parenteral fluids as indicated:
   lactated ringers and 2.5% Dextrose
   IV—brachial vein
   SQ—inguinal (leg) or dorsal cervical (neck)
   at 15–30 cc/kg

7. Entubation:
   15–25 cc/kg lactated ringers and 2.5% dextrose w/2–3 cc/kg Pepto-Bismol at 100° F

8. Quiet, ventilated stabilization area away from drafts, humans, noise, with free access to heat lamps.

Birds showing signs of clinical shock
   Begin with step 6
   Do not administer step 7
   Discretionary: Dexamethasone at 3 mg / kg. SQ (subcutaneous).

   Allow one hour for rest
   Transport as soon as possible thereafter.

samples are collected in microhematocrit tubes to determine packed cell volume and total solids; these diagnostic tests can indicate dehydration, anemia, or possible infections.

A number of oiled birds are dehydrated as a result of enteritis (inflammation of the lining of the gastrointestinal tract). Dehydration is a serious condition resulting in an actual reduction in circulating blood volume. Dehydration can result in compromised kidney function and subsequent permanent kidney damage, with the ultimate sequellae of acidosis and clinical shock.

For a severely dehydrated bird, aggressive fluid therapy is instituted. When blood circulation is significantly reduced or the lining of the intestinal wall is too damaged to take up orally administered fluids, intravenous fluids are administered by bolus injection into the cutaneous ulnar (wing) or medial metatarsal (leg) veins at a rate of 50 to 120 cc/kg per day, divided into 3 daily doses (Welte 1991). At this point of the treatment, dehydration may be the most life-threatening problem the animal faces. Only with aggressive fluid therapy can blood volume and circulation be increased and normal function of lungs, heart, and all basic metabolic activity down to the cellular level be restored.

For less debilitated animals, oral supplementation by gavage (a tube inserted into the crop) with a warm electrolyte solution (Lactated Ringers and 2.5% Dextrose, Pedialyte, etc.) serves to rehydrate the bird while flushing the ingested oil from the gut. An enteric coating agent (Pepto-Bismol is then administered orally by stomach tube to assist in healing and to prevent further absorption of toxic components of the oil.

Heavily oiled birds are not allowed free access to water and food because their oiled bodies frequently contaminate their food sources and drinking water. Rehydrating fluids are given to these birds by gavage. Nutrients are later added to the tubing solution, and tube-feeding is repeated every 4 to 6 hours until the bird is cleaned and is then permitted free access to food and water.

After rehydration the bird is kept warm and quiet, away from people and other stressors. It is not washed until it is alert, responsive, and restored to normal fluid balance. Efforts are made to stabilize the bird and wash it within 24 hours.

Rapid retrieval, stabilization, and cleaning of oil-soaked birds is vital to successful treatment; many refined petroleum products possess components that can be transcutaneously absorbed, intensifying internal problems.

A captive injured wild animal reacts to captivity differently than a domestic pet does, making wild animals more difficult to treat. Wild animals view man as a predator. An oiled bird does not recognize the good intentions of its caregivers; it is in fear for its life throughout its treatment. The wild animal reacts to rehabilitation efforts in a pattern of response familiar to wildlife rehabilitators and wildlife veterinarians: The animal produces elevated levels of corticosteroids (such as norepinephrine or adrenaline). These naturally produced steroids assist the animal in coping with stressful situations but also decrease its immune system defenses, rendering the animal vulnerable to secondary diseases.

This is not the only problem inherent in wildlife rehabilitation. A wild animal masks all symptoms, making it difficult for caregivers to visually evaluate the animal's true condition. Many wild animals do not recognize food in captivity and will not eat. Most have never been confined and can injure themselves trying to get out of a cage. And in attempting to defend themselves, wild animals can cause serious harm to human handlers. All treatment methods must be geared to reducing the stress that the animal is exposed to in captivity; to protecting the handler; and to incorporating special techniques for housing, feeding, and evaluating wild animals. With oil spills, the challenge is intensified, when twenty,

Figure 5.1. Wash lines operated sixteen hours a day early in the spill response.

fifty, or one hundred injured wild animals are delivered at once and each animal is in need of immediate care.

Removing Oil from the Feathers

Since a bird's abilities to fly and to remain waterproof are dependent on the interlocking structure of feather barbs, barbules, and barbicels, oil must be removed from the feathers without damaging the delicate feather structure (Figs. 5.1 and 5.2).

Oiled birds cannot be washed unless large amounts of hot water are available. To wash one duck 100 hundred gallons of 103 to 105°F water are needed over a 20-minute period. This volume of hot water can be assured only with industrial hot-water heaters or a series of steam generators. Water over 105°F can harm the bird, but the water must be above 102°F to lift the oil. Anyone who has tried to wash a greasy frying pan in cold water will understand the necessity for very warm water.

Numerous cleaning agents have been tested for their ability to remove oil from feathers (Frink 1987; Bryndza 1991). The cleaning agent must be nonirritating to animals and humans; it must be able to lift the oil and

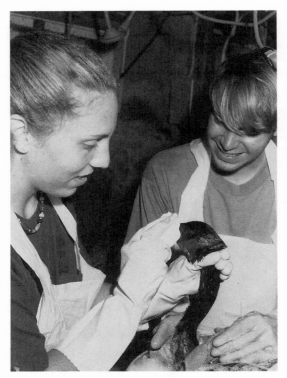

Figure 5.2a. Initial washing of a Canada goose.

maintain it in suspension. The cleaning agent must be rinsed quickly and completely from the feathers. Very few cleaning agents meet the criteria. Dawn dishwashing detergent (Procter & Gamble) has performed best in all tests. Only previously tested and accepted detergents should be used on oiled animals; no experimentation should be done during an oil spill response.

Each bird is cleaned by a team of three or four trained people in a warm, quiet area, free from drafts. The actual cleaning process involves patient, gentle cleansing of the feathers without disrupting the delicate feather structure. A bird may be moved through three or more 10-gallon tubs of warm water with a 2 to 10 percent concentration of detergent.

After all of the oil is removed from the feathers, the bird is *completely* rinsed with clean, warm water. Since detergents are hydrophylic, any soapy residue can impede waterproofing. Rinsing is carried out with a combination of spray rinses and tubs of clean water at 103 to 105°F. The bird is not accceptably rinsed until *diamondlike beads of water roll*

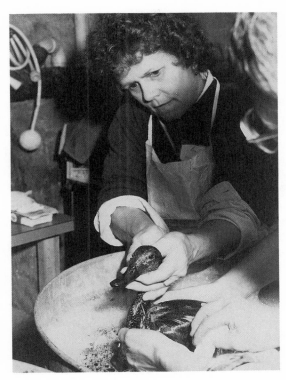

Figure 5.2b. Final wash of a female mallard.

freely from the feathers. This is the one sign of a successfully cleaned oiled bird and is recognized as the end point in every oiled-bird cleaning effort. The failure to rinse the bird adequately is probably the most common cause of unsuccessful rehabilitation efforts.

The newly washed bird is placed in a clean holding pen to dry. The pen is lined with sheets or towels and curtained to minimize human intrusion. Free access to (and away from) heat lamps allows the bird to find a comfortable ambient temperature.

Once the bird has been stabilized and cleaned, it is allowed free access to water and a variety of foods. The staff monitors the birds to see which ones are self-feeding. The droppings are checked for blood and oil, and these birds are treated for enteritis and fed easily absorbed nutrients by stomach tube.

Twenty-four hours after being cleaned, the birds are allowed to swim. They need free access into and out of large amounts of water, where they can actively swim and preen. Waterbirds usually take to the water readily; when they begin to get wet they leave the water to preen. As a bird continues its efforts to swim then preen, it realigns its feathers and restores original feather structure. This alignment of the feathers is what ensures the bird's waterproofing. The feather structure does not require, but is further enhanced by, the application of preen-gland wax from the bird's own uropygial gland. This natural oil seems to assist in maintaining the feather alignment, much as hair spray holds a hairstyle (Ridjke 1970). Birds that are waterproof will demonstrate the diamond beading of water on their feathers. They will also be able to remain in the water without getting waterlogged.

At the rehabilitation center, different housing, pools, and general care are provided according to species. A common loon, a great-blue heron, and a western sandpiper each has special husbandry and treatment needs that an experienced wildlife rehabilitation staff can address during the spill response.

Criteria for Release

The rehabilitated oiled bird should be of average weight for its species and sex to be a candidate for release. It should be adequately muscled so that it can forage normally in the wild. The bird must be examined by experienced wildlife veterinarians; birds with any signs of disease or debilitation should not be released into wild populations. The healthy and waterproof bird is slowly exposed to temperatures comparable to outside weather. This, of course, is critical in the cold winter months.

Birds are banded with Fish and Wildlife Service bands and released early in the day in appropriate, and oil-free, habitat.

Wildlife Rehabilitation Needs in Major Spills

The preceding material described the rehabilitation of a *single* oiled bird. Oil spills that contaminate a large number of birds present a real challenge. The intense human labor, the immediacy of the effort, and the need for special equipment (water, detergent, pools, medication, etc.) require significant resources and an experienced team to manage the resources and the people. A spill involving one hundred or more birds taxes the resources of even those of us who like to think we are always prepared for oil spills.

For example, a facility receiving fifty Canada geese a day needs an area of 800 square feet for initial housing of the oiled birds; a second large, enclosed area for washing the birds; a way to produce over 6,000 gallons of 103°F water in 1 day, and provisions for acceptable disposal of 1,000 gallons of oily wastewater daily.

Staff must be Occupational Safety and Health Administration–trained, and all volunteers must receive complete right-to-know information.

The 1990 Exxon Spill in the Arthur Kill

Setup of the Wildlife Treatment Program

Following the Exxon oil spill, Tri-State sent a three-person assessment team to the spill site. The team's goals included advising the spiller and government on deterrent techniques to prevent wildlife from becoming oiled, assembling and training personnel to retrieve oiled animals, and confirming that state and federal laws were understood and followed.

Normally, when oiled birds are retrieved, the assessment team calls in a response team, which mobilizes and staffs field stations for treatment of contaminated animals. On their arrival at the Bayway Refinery, however, the assessment team discovered that birds had already become contaminated by no. 2 diesel fuel. Two additional Tri-State staff were called in, and the assessment team at this time also took on the responsibilities of a response team.

1. On-site training was provided for cleanup contractors and Exxon personnel who would work in the field to retrieve oiled animals.

2. Training sessions were conducted for refinery personnel who would assist with stabilization of animals delivered to a predesignated field stabilization center.

3. A field stabilization area was set up at a trailer on the refinery grounds and birds were stabilized (see Table 5.1) before being transported by closed van to the Tri-State wildlife rehabilitation facility in Newark, Delaware. Vans made the 2-hour trip one to three times a day.

In Newark, the treatment facility received its first animals the evening of January 3. Staff operated the facility, and experienced volunteers were recruited to assist in treating the affected animals.

Species delivered alive to the rehabilitation facility included herring gull (53), hooded merganser (1), great black-backed gull (27), mallard (12), Canada goose (15), black duck (7), gadwall (2), domestic goose (1), ruddy duck (1), ring-billed gull (2), double-crested cormorant (1), bufflehead (1), and greater scaup (1).

Examination and Treatment

Medical-treatment protocols used were those previously discussed. The birds that arrived early in the spill had been severely affected by the volatile components of the diesel fuel and required intensive care. Examination of living and dead birds delivered within the first week following the oil spill disclosed severe respiratory distress due to irritation, inflammation, or hemorrhaging of the lungs and air sacs. One duck, foraging for food close to the pipeline leak, died from acute respiratory distress so quickly that food was still in its mouth when it died. Postmortem examination indicated severe pulmonary edema and hemorrhaging. Mortality was highest in birds retrieved during the first 72 hours after the spill.

All of the birds delivered during the first few days following the spill were moderately to severely contaminated by oil. Oil was present in the mouths and eyes and in the droppings of many of the birds. Highly aggressive supportive care was provided for the first arrivals. The intensive nursing care required considerable investment of time from skilled Tri-State staff and trained volunteers.

Gulls were the most numerous taxon delivered to Tri-State for treatment and also the most numerous taxon retrieved dead and alive from the marshes. Herring and great black-backed gulls are large, aggressive birds with a wing span of over 5½ feet. These larger gulls always pose a threat to handlers; this and the technical aspects of the medical care meant that only staff and trained, experienced volunteers could be used to perform treatments. Fluids were provided parenterally (intravenously or subcutaneously) and *per os* (orally, by stomach tube) three times a day to each bird to rehydrate the animals and to counteract kidney failure.

This meant that during the peak-response period (days 2 through 9), three teams of three people tubed up to one hundred birds during each of the three shifts. In other words, twenty-seven people were required each day just to provide medical care. Additional personnel were needed to clean the oiled animals, schedule volunteers, and operate the facility.

The fifty-one birds delivered over the first 4 days of the response presented all or some of the following physical and clinical signs:

hypothermia or reduced body temperature

oil, moderate to severe, on feathers and skin

inflammation of conjunctiva and corneal surface of the eyes

oil in mouth, nares, vent

acute respiratory distress

tarry black (bloody or oiled) or green (bile-stained) droppings sternal recumbency (inability to stand)

ataxia (weakness or incoordination)

Some of the seventy birds delivered later in the spill were moderately to severely underweight and dehydrated. Basic bloodwork was done on some birds; complete blood counts were done on a select number of the birds.

Cleaning and Aftercare

After the birds were stabilized, they were washed in a series of three 10-gallon tubs of water at 104°F with a 3 to 10 percent solution of Dawn detergent, using the protocol described earlier.

Normally after they are washed and dried, the cleaned birds are allowed access to fresh food and to swimming pools holding 40 gallons (dabbling ducks) to 800 gallons (common loons and diving ducks) of water. The birds in the Arthur Kill spill were too sick to enter this stage of the rehabilitation process. Because of severe debilitation, most of the birds needed nursing care for an extended period of time. Many of the gulls, which normally eat readily after cleaning, did not self-feed for over 3 weeks. This could be attributed to kidney damage and neurological impairment of unknown etiology.

The treatment teams caring for the Arthur Kill oiled wildlife operated from 8:00 A.M. to 2:00 A.M. for the first 10 days following the spill. The hours were reduced to 8:00 A.M. to midnight by day 15. The teams remained active for over 6 weeks, due to the extended period of time that Exxon was required to monitor the river and adjacent tributaries.

None of the birds received during the last 10 days of the response (5 to 6 weeks postspill) were oiled. The wild birds being delivered to the facility during this time presented the types of injuries seen in birds delivered to mid-Atlantic rehabilitation facilities during January and February: traumatic injuries, including wing and leg fractures; end-of-winter debilitation, and weight loss that could be caused by a number of disease entities.

The Exxon field retrieval and shuttle crews, and the environmental coordinators were all exceptionally responsive to Tri-State's requests and comments and were cooperative in every way with our work. At the peak of response, Exxon had five boats with fifteen crew in the river and marshes each day. Twenty-five Exxon personnel were provided to help with wildlife retrieval, on-site stabilization, and transport. The Delaware facility admitted 124 for treatment. Release rates varied considerably depending on some unusual clinical findings, discussed next. Caseload disposition is presented in Table 5.2.

Table 5.2. Caseload Disposition of Oiled Birds

Caseload	Oiled	Unoiled
Oiled Only		
Released	41	
Died	21	
Euthanized	9	
Neurologic		
Released	8	8
Died	12	8
Euthanized	2	5
Injury/disease		
Released	1	3
Died		6
Euthanized	1	1

Caseload Disposition

Fifteen oiled birds retrieved in the field died during transport. Sixteen oiled birds were euthanized or died at the field stabilization site. One hundred twenty-four birds were admitted to the Delaware facility for treatment. Fifty-eight percent of oiled birds (not displaying neurological signs) were released after treatment; only 26 percent of the oiled birds displaying neurological impairment were released after treatment (see Table 5.2). Other birds from the spill were handled by NYDEC.

Unusual Findings

A significant number of oiled and unoiled birds delivered to the facility displayed neurological signs not consistent with oil contamination (Fig. 5.3). These birds were uncoordinated, ataxic, and anorexic. Most of them were unable to stand. Because of the unusual nature of the clinical signs, additional tests were performed on a selected number of the birds. Gas chromatography of the oil was done, complete blood panels were done on certain birds, and blood was analyzed for lead levels; results were nondiagnostic. Unoiled birds delivered later in the spill demonstrated the same debilitation and neurological impairment. Blood was taken on February 1 for analysis for avian influenza; test results were negative.

Dr. Virginia Pierce, director of pathology at the Philadelphia Zoological Gardens, was contracted to perform postmortems on the birds delivered dead from the spill site or dying at the center. She performed complete basic postmortem examinations and prepared duplicate pathology slides for Exxon. Pierce's initial findings (forty-five birds examined) confirmed the typical oil spill toxicosis syndrome of kidney/gastrointestinal/respiratory involvement.

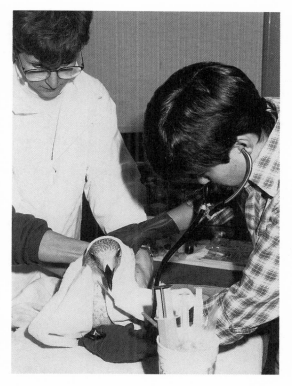

Figure 5.3. Dr. Heide Stout treats a gull showing neurologic signs.

Pierce noted the presence of *Aspergillus fumigatus* in a significant portion of the gulls necropsied. Aspergillosis is a fungal infection that is extremely resistant to treatment. *Aspergillus* forms colonies of bluish, pale yellow to creamy-white fuzzy or cheesy plaques within the body cavity. This fungal infection is frequently found on the lungs or air sacs but can disseminate quickly throughout the body, occluding the trachea and surrounding the heart, liver, or kidneys.

Aspergillus is a ubiquitous, opportunistic fungus that can spread through the air in the wild or within rehabilitation centers if animals are bedded in moist straw or organic materials or left in damp, dark areas. Pierce's findings in the Arthur Kill spill indicate that many birds retrieved after the Arthur Kill spill already had advanced *Aspergillus* growths in the wild: "Of the gulls which were euthanized after treatment at Tri-State, the one ring-billed gull and seven herring gulls which had aspergillosis most likely had the infections upon presentation at the rehabilitation center, as these birds were at the center for less than seven days and in all cases had severe multifocal chronic aspergillosis airsac-culitis and pneumonitis. Ten great black-backed gulls had aspergillosis, varying from minimal focal airsacculitis to severe multifocal chronic aspergillosis airsacculitis and pneumonitis. These birds were treated at the rehabilitation center for variable periods ranging from two weeks to three months" (Pierce 1991). It was of great interest to study the birds delivered from the wild with signs of severe clinical disease; however, no identification has been made for the causative agents responsible for the neurological impairment and advanced fungal infections of the birds retrieved following the Arthur Kill spill.

Heavy infestations of *Aspergillus fumigatus* continue to be docu-mented in populations of gulls affected by the Exxon oil spill, a fact that has prompted a study, begun by the U.S. Fish and Wildlife Service in 1992, to determine the prevalence and distribution of this epornitic disease.

Problems and Concerns with Rehabilitation Protocols

Because of the number of agencies concerned with the oiled wild birds, the caseload records became increasingly difficult to monitor. Exxon assigned field numbers to each bird; Tri-State assigned its own case numbers and identifying leg tags to each bird; Pierce, the pathologist, assigned numbers to birds as did the pathologist for the state of New York. Some birds were identified by four or five different numbers. Confusion in records did not affect the rehabilitative care of the birds but did preclude certain types of statistical studies.

Another problem arose in the lack of consensus between numerous wildlife agencies representing federal wildlife officials, two states, mu-nicipal government agencies, and industry representatives. The agen-cies differed in opinion on who should be responsible for recovery of the animals, how long the rescue efforts should be continued, where the search and rescue should take place, and what reports should be pre-pared for which government agencies. This was complicated by disagree-ment on which agency had final authority to make these decisions. This lack of agreement did not cause problems for Tri-State except in its efforts to comply with all federal, state, and municipal requests and regulations while providing rapid care for the affected animals.

Because of excellent state–federal–industry contingency planning in the Delaware Bay–River system, Tri-State is accustomed to more smoothly coordinated efforts in oil spill responses. In our region, all government and private agencies that are activated during an oil spill meet every year to review response plans and refine protocols.

Conclusions

Tri-State Bird Rescue & Research was called in by the U.S. Fish and Wildlife Service and contracted with by Exxon for wildlife rehabilitation following the pipeline spill in the Arthur Kill. Tri-state staff operated an on-scene stabilization site at Exxon's Bayway Refinery, and a rehabilitation facility in Delaware to treat affected animals for a 6-week period beginning January 2, 1990.

One hundred twenty-four live birds were delivered to the wildlife rehabilitation facility. One-third of the birds delivered oiled and unoiled presented neurological signs of unknown etiology and not consistent with oil contamination. A significant number of the birds and carcasses retrieved following the Arthur Kill spill had advanced growths of *Aspergillus fumigatus*, determined to be present before oil contamination occurred.

Rehabilitation of the oiled birds followed the basic protocols of medical treatment and cleaning. Fifty-eight percent of the oiled birds not displaying neurological impairment were treated and released; 26 percent of the oiled birds with neurological problems recovered and were released.

Acknowledgments

I thank Heidi Stout, V.M.D., for valuable comments on the manuscript. Chris Hileman and Joyce Ponsell provided caseload information and data analysis for this chapter.

Literature Cited

Albers, P. H. 1977. Effects of external applications of fuel oil on hatchability of mallard eggs. In D. A. Wolfe, ed., *Fate and Effects of Petroleum Hydrocarbons in Marine Ecosystems and Organisms*, pp. 158–163. New York: Pergamon Press.

————. 1991. Oil Spills and the environment: A review of chemical fate and biological effects of petroleum. In J. White and L. Frink, eds., *The Effects of Oil on Wildlife: Research, Rehabilitation and General Concerns*, pp. 1–12. San Suisun: International Wildlife Rehabilitation Council.

Bryndza, H. E. 1991. Surfactant efficacy in removal of petrochemicals from feathers. In White and Frink, *Effects of Oil*, pp. 78–94.

Dein, F. J., and L. S. Frink. 1986. Rehabilitation of oil-contaminated birds. In

R. W. Kirk, ed., *Current Veterinary Therapy*, pp. 719–723. Philadelphia, Pa.: Saunders.

Frink, L. 1987. An overview: Rehabilitation of oil contaminated Birds. Proceedings of the 1987 Conference on Prevention and Control of Oil Pollution, American Petroleum Institute, Washington, D.C.

Langenberg, J., and F. J. Dein. 1983. Pathology of ruddy ducks (*Oxyura jamaicensis*) contaminated with spilled #6 fuel oil. In D. Rosie and S. N. Barnes, eds., *The Effects of Oil on Birds: A Multi-discipline Symposium*, pp. 139–142. Newark, Del.: Tri-State Bird Rescue & Research.

Leighton, F. A. 1991. The toxicity of petroleum oils to birds: An overview. In White and Frink, *Effects of Oil*, pp. 43–57.

Pierce, V. 1991. Pathology of wildlife following a #2 fuel oil spill. In White and Frink, *Effects of Oil*, pp. 78–94.

Ridjke, A. M. 1970. Wettability and phylogenetic development of feather structure in water birds. *J. Expt. Biol.* 52:469–479.

Welte, S. 1991. *Oiled Bird Rehabilitation: A Guide for Establishing and Operating a Treatment Facility for Oiled Birds.* Newark, Del.: Tri-State Bird Rescue & Research.

White, J. 1991. *Tenyo Maru* spill. *Wildlife and Oil Spills Bulletin* 2:1.

6. Bioremediation and the Arthur Kill

Joanna Burger

All segments of society are concerned with degradation of environmental quality, as it affects both human health and ecosystem health. Thus, there is public and governmental pressure to reduce the incidence of industrial accidents, including those involved with the transport and storage of chemicals. Furthermore, when accidents do occur, a high premium is placed on immediate cleanup, restoration, and rehabilitation. This is true with oil because it can so easily spread, particularly on a water medium, and has potential long-term effects.

Methods of cleaning up oil following major spills include physically picking up and collecting the oil as well as reducing the amount of oil by degrading it with the use of microbes. Microorganisms can convert petroleum hydrocarbons to carbon dioxide and water, thus effectively eliminating the oil. The end products of biodegradation, water and carbon dioxide, are nontoxic and do not harm the environment or living organisms. The use of microorganisms to eliminate oil is also called "bioremediation," because biological agents (microbes) are used to remedy the situation.

Following the 1990 Exxon oil spill in the Arthur Kill, Exxon mounted a massive cleanup operation, directed by the Coast Guard, that included extensive physical cleanup procedures (skimming and picking up oil from the water's surface). Exxon also conducted a small-scale bioremediation experiment on Pralls Island. Because the results of this experiment are important for decisions concerning the most appropriate responses to future oil spills, I summarize them here based on a report to Exxon by scientists from Exxon Research and Engineering Company and DuPont Environmental Remediation Services, Inc. (Madden, Hinton, and Lee 1991).

This chapter describes the history of hydrocarbon biodegradation and oil spill remediation, examines the bioremediation experiment conducted on a contaminated beach on Pralls Island, and discusses how communities can be prepared to use bioremediation for oil spills in the future. It also describes briefly the factors that influence rates of petroleum biodegradation.

History of Hydrocarbon Biodegradation and Oil Spill Bioremediation

Early work during the 1950s and 1960s on microbial use of petroleum hydrocarbons was performed with the objective of producing microbial biomass for food to prevent world food shortages (Champagnat and Llewelyn 1962; Champagnat 1964; Atlas and Bartha 1992). Hydrocarbon sources (petroleum) were thought of merely as an inexpensive carbon source for this food production methodology. At that time, obviously, oil was not considered a high-cost, scarce resource. Basic research involved examining the metabolic pathways of hydrocarbon use by different microorganisms (Foster 1962; National Academy of Sciences 1975; Higgins and Gilbert 1978).

By the late 1960s, people began to realize that oil reserves were limited (and were nonrenewable), and there would inevitably be fuel shortages (Atlas and Bartha 1992). Indeed, by the 1970s the world faced a crisis in fuel energy resources, and it became clear that petroleum was uneconomical as a substrate for producing microbial biomass.

The dependence of developed countries on the oil resources of the Middle East resulted in large-scale movement of oil about the world in large ocean tankers. A series of wrecks of large oil tankers, resulting in massive oceanic oil spills, focused world attention on degradation of the oceans by oil (National Academy of Sciences 1975, 1985). Scientists gradually shifted their research to using microbes to biodegrade oil under real environmental conditions. Instead of using petroleum to produce microbial biomass, the aim became to use microbes to clean up petroleum pollutants. Experiments were conducted to understand the factors that limit the rate of petroleum breakdown by microbes in nature.

Many of these early, classic experiments were performed by Atlas (1975, 1977, 1979, 1981, 1984) and Atlas and Bartha (1972a, 1972b, 1972c, 1973a, 1973b, 1973c, 1973d) at Rutgers University. They conducted their studies in the laboratory and in the field, as well as on actual oil spills. These studies showed that the fate of oil in nature is determined by the population size and structure of the microorganisms that can degrade oil and by the abiotic factors that control the growth of these microbial populations. The rate of degradation of oil in the environment is limited primarily by temperature and nutrients (nitrogen, phosphorus, and oxygen) (Bartha and Atlas 1973; Atlas 1981, 1984; Bartha 1986; Leahy and Colwell 1990).

As might be expected given the magnitude of oil transport and storage, the importance of oil as a fuel source in the world economy, and the potential for environmental contamination from oil spill accidents, a great deal of research is being performed on the factors affecting the rates that microorganisms can break down oil under different environmental conditions. Whether conducted in the laboratory or in the field, experiments are still experiments, and the real test for the usefulness of microbes in cleaning up oil spills remains their use following a real oil spill in the estuaries or oceans of the world. It is in this light that the data from the Arthur Kill are useful.

Biodegradation of Petroleum

Biodegradation of petroleum in the environment is a complex process that depends on the nature and amount of oil present, the ambient and seasonal environmental conditions, and the composition of the microbial community present (Cooney 1990; Atlas and Bartha 1992). Petroleum hydrocarbons are soluble in water only at very low concentrations, and oil spills almost always exceed the solubility limits (Harrison et al. 1975). Consequently, the oil spreads across the water surface. The degree of spreading determines the amount of surface area available for attack by microbes. Biodegradation of oil occurs at the oil–water or oil–soil interface; thus, when oil forms a thick layer on either water or soil, the rates of biodegradation are slowed down (see Atlas et al. 1980).

The ability of microorganisms to degrade oil also depends on the composition and population size of the microbial community. Hydrocarbon-degrading bacteria and fungi are widely distributed in marine, freshwater, and soil habitats (Atlas and Bartha 1992). The most prevalent genera of hydrocarbon-using microorganisms in aquatic environments are *Pseudomonas, Achromobacter, Arthrobacter, Micrococcus, Nocardia, Vibrio, Acinetobacter, Brevibacterium, Corynebacterium, Flavobacterium, Candida, Rhodotorula,* and *Sporobolomyces* (Bartha and Atlas 1977). Microorganism communities that are exposed to oil become adapted, experience increased proportions of hydrocarbon-degrading bacteria, and can respond to the presence of oil pollutants within hours (Leahy and Colwell 1990).

Temperature affects the rate at which microorganisms degrade oil because it affects the physical nature of the oil as well as the rates of hydrocarbon metabolism by the microbes (Atlas 1981; Leahy and Colwell 1990). Low temperatures decrease the viscosity of oil and the volatilization of toxics from the oil, decreasing biodegradation (Walker and Colwell 1974; Atlas 1975).

In addition to suitable temperatures, molecular oxygen is required for the initial steps of biodegradation by bacteria and fungi (Atlas 1984). Furthermore, microorganisms require nitrogen, phosphorus, and other nutrients for biodegradation. Atlas and Bartha (1972b) reported that nitrogen and phosphorus in seawater are severely limiting for microbial hydrocarbon degradation, suggesting that their addition could increase the rate of degradation. Bioremediation is necessary because nutrients for biodegradation by microbes can be limited.

Bioremediation as a Cleanup Technique

Bioremediation uses microorganisms to degrade, and thus eliminate, oil pollutants. Since the end products (water, carbon dioxide) are nontoxic, remediation does not harm the environment (Atlas and Bartha 1992). Ecologists may, however, caution the use of bioremediation because of the effects nutrient enrichment might have on other organisms in the system. Although this was an early concern (Atlas and Bartha 1973d; Bartha 1976), oleophilic fertilizers generally do not stimulate algae blooms.

The use of bioremediation is inexpensive compared with the costs of physical methods of cleaning and decontaminating the environment. Unfortunately, no national, coordinated response for oil spill bioremediation exists (Gregorio 1991). A detailed review of laboratory and field experiments with bioremediation and applications of hydrocarbon biodegradation for bioremediation can be found in Atlas and Bartha 1992. Here I summarize only the results from the recent *Exxon Valdez* accident.

The Exxon tanker *Valdez* ran aground in Prince William Sound, Alaska, on March 24, 1989, spilling 11 million gallons of crude oil (Environmental Protection Agency 1989; Kelso and Kendziorek 1991). Although the initial approach to cleanup involved physical methods, bioremediation was subsequently applied. A joint study between Exxon and the U.S. Environmental Protection Agency (EPA) tested three types of nutrient supplementation in the field after laboratory tests (Tabak et al. 1991; Glaser, Venosa, and Opatken 1991). The most effective treatment (an oleophilic fertilizer Inipol EAP 22) was subsequently approved for shoreline treatment of exposed oil. Non oleophilic, slow-release fertilizers were used for treatment in the sediments and under rocks.

Monitoring by a joint Exxon, EPA, and State of Alaska Department of Conservation team indicated that rates of oil biodegradation increased fivefold following fertilizer application (Prince, Clark, and Lindstrom 1990; Pritchard and Costa in press). Furthermore, no adverse ecological effects of bioremediation were shown (Fox 1990). It is in light of the Alaska experience that the bioremediation experiment of a contaminated beach on Pralls Island in the Arthur Kill was initiated in 1990.

Bioremediation on Pralls Island, Arthur Kill

The initial cleanup operations following the January 1–2, 1990, Exxon oil spill involved physical recovery of oil from the water surface and mechanical cleanup of shoreline habitats using sorbent and portable vacuum equipment (see Chapter 1). Physical-recovery operations were necessarily suspended on Pralls Island during the nesting season of the herons, egrets, ibises, and gulls (see Chapter 13). During the breeding season bioremediation was identified as a viable technique for the cleanup of oil remaining on the Pralls Island shoreline. The following description of the bioremediation of Pralls Island is based on the report to Exxon on the bioremediation work (Madden, Hinton, and Lee 1991).

Methods

Bioremediation on Pralls Island was conducted by Exxon Research and Engineering Company with the support of DuPont Environmental Remediation Services. The program was approved by the EPA, National Oceanic and Atmospheric Administration, New York State Department of Environmental Conservation, and New York City Department of Parks and Recreation. The program ran from September 1 to December 3, 1990, and covered a $\frac{3}{8}$-mile section of sand beach on the southwestern shore of the island.

After the decision to use bioremediation was made in March, a site assessment was conducted, and 2,000 feet of shoreline were slated for bioremediation. Core samples were taken in July to map the location and extent of oiling on the beach.

The beach was treated once in early September 1990 with a slow-release fertilizer (Customblen, a product of Sierra Chemicals, Milpitas, California). The dosage was based on Exxon's experience in Prince William Sound following the *Valdez* accident. On Pralls Island a linear dose of 610 grams per meter (0.4 lbs./ft.) was used in each of two trenches, based on an estimated oil zone width of 12 meters (40 ft.). To prevent contamination of any remaining birds or other wildlife, the fertilizer pellets were buried in two shallow trenches 4 to 6 inches below the surface, the depth to which the oil seeped. A small control section of beach was left untreated in the middle of two treatment sections.

The study design required monitoring of inorganic nutrients (nitrogen, phosphorus, dissolved oxygen, ammonia), soil microbiology, total petroleum hydrocarbons, and oil concentrations by gas chromatography. Sampling wells or seepers (N=35) were installed along the beach for examining interstitial water. Five sets of wells were installed in the treated zone between low and high tide and two sets in the untreated zone.

Sediment cores were taken at seeper locations 8, 22, 43, 64, and 92 days following fertilizer treatment. The microbial population was monitored by performing heterotrophic and hydrocarbon-utilizing microbial counts on the soil samples. Microbial activity was measured with biomineralization assays to estimate biodegradation potential. The assays were

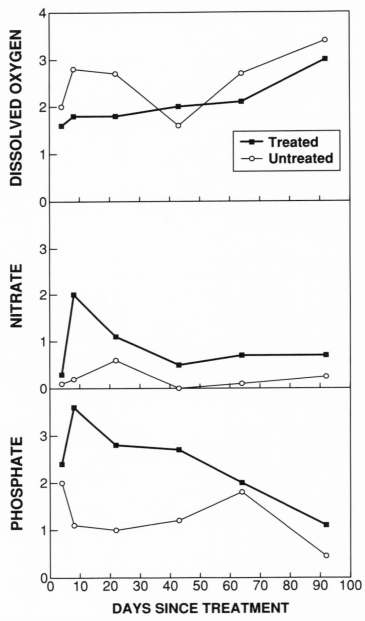

Figure 6.1. Changes in nutrient levels (mg/kg) in the treated (fertilized) and control sites on Pralls Island.

conducted in the laboratory using radioactive tracers (C^{14}) to detect the rate of oxidation of the compounds to $C^{14}O_2$.

Results

The analytical results from the water samples indicated significant differences between the treated and control zones. An increase in the level of nutrients, with a decreased level of dissolved oxygen, was found in the treated compared with the control area (Fig. 6.1). Lower dissolved-oxygen levels reflect the utilization of oxygen by microbes to metabolize oil. The low difference in dissolved oxygen in the two plots was partially a result of some leakage of nutrients from the treated to the control zone.

The increase of nutrients in the treated zone indicates that the application strategy worked on Pralls Island. Thus, even though fertilizer was delivered in trenches rather than evenly over the beach and was buried rather than applied on the surface, nutrients reached the beach. However, a significant concentration gradient existed across the beach, with more total nitrogen near the low-tide areas (Fig. 6.2). Nutrient levels in the near-shore waters did not increase significantly, indicating that eutrophication was avoided.

During the experiment, both heterotrophic bacteria and hydrocarbon users responded to the treatment with significant increases in populations after fertilizer installation (Fig. 6.3). As is to be expected, microbial populations decreased with time as the nutrient levels and oil levels on the beach diminished. That is, as the microbes used the nutrients and oil, there were fewer available nutrients (both from the fertilizer and the oil), they became limiting, and microbial populations decreased.

The microbial populations in both the treated and control areas increased, although those in the treated areas increased faster. The lag is no doubt due to the added time required for the nutrients to leak from the treated to the control site. The design flaw of contiguous control and treatment sites ensured that nutrients would seep into the control site. Future bioremediation tests should make sure sites are separated sufficiently to prevent such seepage.

The measure of success of any bioremediation program is whether oil is degraded. On Pralls Island the total petroleum hydrocarbons remained relatively constant from mid July 1990, when the study sites were designated, until early September, when fertilizer was added. Both sites experienced a slight increase in the amount of oil in mid summer, due either to the effect of small, chronic oil spills or the redistribution of oil by tidal waters.

Within 3 weeks of fertilization, the average concentration of total petroleum hydrocarbons had dropped to below 400 milligrams per kilo-

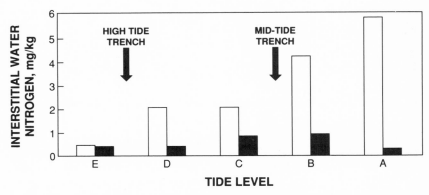

Figure 6.2. Concentration gradient of nutrients across the beach, Pralls Island, follow-ing fertilization. E = high tide, A = low tide mark. Open bars = soil values, solid bars = near shore water.

gram in both sites (Fig. 6.4). This dramatic reduction, particularly in the treated site, was accompanied by increases in nutrient levels on the beach and decreases in dissolved oxygen, indirect measures of increased microbial activity.

Oil concentration remained constant until November, when it in-creased (between days 64 and 92). One possible explanation for the increase is that the beach experienced another oil spill. In fact, a ship (the *Centaur*) spilled 500 gallons of no. 6 oil on Staten Island, just north of Pralls Island, on November 8. This oil may ultimately have been moved to Pralls Island by wind and tides.

Discussion

Madden et al. (1991) concluded that the Pralls Island project demon-strated that bioremediation can be safely and effectively applied in intertidal channels. The addition of fertilizers accelerated the rate of bi-odegradation. The evidence clearly documents that the observed reduc-tion in oil levels on the Pralls Island beach was due to remediation, even though the control site was not sufficiently removed from the treated site to provide clear comparisons.

In my estimation the use of bioremediation on Pralls Island was impor-tant for the following reasons: (1) It provided a demonstration for federal, state, and local agencies to evaluate; (2) it provided a demonstration of land-based bioremediation in a small, intertidal region; (3) it demonstrated that fertilization could be effective even when the material was buried to pre-vent adverse affects to resident birds and other wildlife; and (4) it demon-strated that constraints, such as application in lines rather than over the

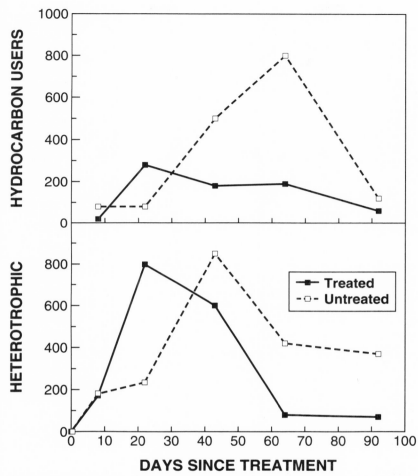

Figure 6.3. Changes in population levels of heterotrophic bacteria and hydrocarbon utilizers (number per ml).

entire area, do not necessarily invalidate the procedures. Most important, of course, bioremediation decreased the oil on the Pralls Island beach.

Conclusions

One important aspect of the bioremediation project on Pralls Island is the valuable lessons learned, both about bioremediation and about how to study bioremediation. The Pralls Island project clearly demonstrated

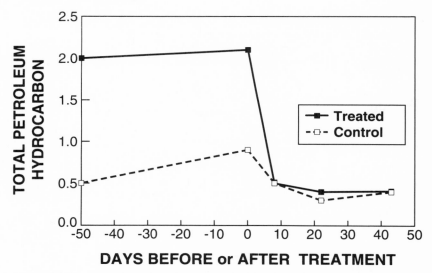

Figure 6.4. Concentrations of total petroleum hydrocarbons (mg/kg) before and after fertilization.

that fertilization can lead to an immediate increase in microbial activity and population levels, with a concomitant decrease in oil. The results demonstrate that even in an imperfect world, where considerations of wildlife health and safety reduce accessibility and constrain methods (i.e., fertilizer had to be buried and could be applied only once), fertilization can enhance and accelerate biodegradation of oil by microbes. More remarkably, fertilization increased biodegradation even though the methods were initiated 8 months after the oil spill, when oil levels on the beach had remained relatively constant for several months (see Fig. 6.4).

The study itself could have been better designed. This, of course, is almost always true, and is always clearer at the end of any study. Nonetheless, the following factors should be kept in mind when designing future bioremediation work:

1. Control sites should be sufficiently separated geographically from treatment sites to prevent any seepage of fertilizer (whether because of proximity or tides).

2. Control sites should, whenever possible, have similar physical conditions, as well as similar initial levels of exposure, as treatment sites. In the Arthur Kill site the control area had only a quarter as much oil as the treatment areas did, making interpretations difficult.

3. Multiple applications of the fertilizer are best to maintain high populations of petroleum-degrading microorganisms.

4. An even application of fertilizer over the impacted area is preferred to the use of linear application.

5. Application of fertilizer should be made as soon as possible after the oil spill to eliminate further dispersion of the oil to other sites via tidal action.

As with all environmental hazards, timing plays a key role: time required to decide to use bioremediation; time required to design the bioremediation program; time required to assemble personnel, equipment, and supplies for the project; and time required to implement bioremediation. All of these times can be reduced by community awareness and preparedness before an oil spill occurs.

On a national scale, the formation of a national bioremediation response team would eliminate the need for duplication of expertise and equipment in countless communities. Even so, communities would be well advised to understand the basic rudiments of bioremediation, have background data on sensitive species and ecosystems in their care, and know the people to contact for expertise on bioremediation should the need arise.

Acknowledgments

I thank the appropriate state agencies and Exxon for making available to me necessary reports on the bioremediation studies, and Richard Bartha for providing important background information on bioremediation and helpful comments on the manuscript. I also thank Michael Gochfeld for comments on the manuscript.

Literature Cited

Atlas, R. M. 1975. Effects of temperature and crude oil composition on petroleum biodegradation. *Appl. Microbiol.* 30:396–403.

———. 1977. Stimulated petroleum biodegradation. *Crit. Rev. Microbiol.* 5:371–386.

———. 1979. Measurement of hydrocarbon biodegradation potentials and enumeration of hydrocarbon-utilizing microorganisms using carbon-14 hydrocarbon-spiked crude oil. Special technical report from American Society for Testing and Materials, Philadelphia, Pa.

————. 1981. Microbial degradation of petroleum hydrocarbons: An environmental perspective. *Microbiol. Rev.* 45:180–209.

————, ed. 1984. *Petroleum Microbiology.* New York: Macmillan.

Atlas, R. M., and R. Bartha. 1972a. Biodegradation of petroleum in seawater at low temperatures. *Can. J Microbiol.* 18:1851–1855.

————. 1972b. Degradation and mineralization of petroleum by two bacteria isolated from coastal water. *Biotechnol. Bioeng.* 14:297–308.

————. 1972c. Degradation and mineralization of petroleum in seawater: Limitation by nitrogen and phosphorus. *Biotechnol. Bioeng.* 14:309–317.

————. 1973a. Abundance, distribution, and oil biodegradation potential of microorganisms in Raritan Bay. *Environ. Pollut.* 4:291–300.

————. 1973b. Effects of some commercial oil herders, dispersants and bacterial inocula on biodegradation of oil in seawater. In D. G. Ahearn and S. P. Meyers, eds., *The Microbial Degradation of Oil Pollutants*, pp. 283–289. Publication no. LSU-SG-73-01. Baton Rouge: Center for Wetland Resources, Louisiana State University.

————. 1973c. Fate and effects of oil pollution in the marine environment. *Residue Rev.* 49:49–85.

————. 1973d. Stimulated biodegradation of oil slicks using oleophilic fertilizers. *Environ. Sci. Technol.* 7:538–541.

————. 1992. Hydrocarbon biodegradation and oil spill bioremediation. *Advances in Microbial Ecology* 12:287–338.

Atlas, R. M., G. Roubal, A. Bronner, and J. Haines. 1980. Microbial degradation of hydrocarbons in mouse from IXTOC-I Cruises. In *Proceedings of Conference on Researcher/Pierce IXTOC-I Cruises*, pp. 1–24. Miami, Fla.: National Oceanographic and Atmospheric Administration, Atlantic Oceanographic and Meteorological Laboratories.

Bartha, R. 1976 (May 25). Biodegradation of oil on water surfaces. U.S. patent no. 3,959,127. May 25 1976.

————. 1986. Biotechnology of petroleum pollutant biodegradation. *Microb. Ecol.* 12:155–172.

Bartha, R., and R. M. Atlas. 1973. Biodegradation of oil in seawater: Limiting factors and artificial stimulation. In Ahearn and Meyers, eds., *Microbial Degradation of Oil Pollutants*, pp. 147–152.

————. 1977. The microbiology of aquatic oil spills. *Adv. Appl. Microbiol.* 22:225–266.

Champagnat, A. 1964. Proteins from petroleum fermentations: A new source of food. *Impact* 14:119–133.

Champagnat, A., and D.A.B. Llewelyn. 1962. Protein from petroleum. *New Sci.* 16:612–613.

Cooney, J. J. 1990 (September 17–18). Microbial ecology and hydrocarbon degradation. Paper presented at the Alaska Story Symposium, Cincinnati, Ohio.

Environmental Protection Agency. 1989. Alaskan oil spill bioremediation project, EPA 160018-891073.

Foster, J. W. 1962. Hydrocarbons as substrates for microorganisms. *Antonie van Leeuwenhoek; J. Microbiol. Serol.* 28:241–274.

Fox, J. E. 1990. More confidence about degrading work. *Bio/Technology* 8:604.

Glaser, J. A., A. D. Venosa, and E. J. Opatken. 1991. Development and evaluation of application techniques for delivery of nutrients to contaminated shoreline in Prince William Sound. *Proceedings of the 1991 International Oil Spill Conference:* 559–562.

Gregorio, F. M. 1991 (January). Key environmental factors to consider in the biotreatment of oil spills. *Genetic Eng. News:*3–4.

Harrison, W., M. A. Winnik, P.T.U. Kwong, and D. Mackay. 1975. Crude oil spills: Disappearance of aromatic and aliphatic components from small sea-surface slicks. *Environ. Sci. Technol.* 9:231–234.

Higgins, I. J., and P.D. Gilbert. 1978. The biodegradation of hydrocarbons. In K.W.A. Chater and H. J. Somerville, eds., *The Oil Industry and Microbial Ecosystems,* pp. 80–117. London: Hayden.

Kelso, D. D., and M. Kendziorek. 1991. Alaska's response to the *Exxon Valdez* oil spill. *Environ. Sci. Technol.* 25:16–23.

Leahy, J. G., and R.R. Colwell. 1990. Microbial degradation of hydrocarbons in the environment. *Microbiol. Rev.* 54(3):305–315.

Madden, P. C., S. M. Hinton, and M. D. Lee. 1991. Bioremediation of a contaminated beach on Pralls Island. Report to Exxon and New Jersey Department of Environmental Protection and Energy.

National Academy of Sciences. 1975. *Petroleum in the Marine Environment.* Washington, D.C.: National Academy Press.

———. 1985. *Oil in the Sea: Inputs, Fates, and Effects.* Washington, D.C.: National Academy Press.

Prince, R., J. R. Clark, and J. E. Lindstrom. 1990. Bioremediation monitoring program. Joint report of Exxon, the U.S. Environmental Protection Agency, and the Alaskan Department of Environmental Conservation, Anchorage, Alaska.

Pritchard, H. P., and C. F. Costa. In press. EPA's Alaska oil spill bioremediation report. *Env. Sci. Technol.*

Tabak, H. H., J. R. Haines, A. D. Venosa, J. A. Gloser, S. Desai, and W. Nisamaneepong. 1991. Enhancement of biodegradation of Alaskan weathered crude oil components by indigenous microbiota with the use of fertilizers and nutrients. *Proceedings of the 1991 International Oil Spill Conference:* 583–590.

Walker, J. D., and R. R. Colwell. 1974. Microbial degradation of model petroleum at low temperatures. *Microbiol. Ecol.* 1:63–95.

Biological Effects

7. Immediate Effects of Oil Spills on Organisms in the Arthur Kill

Joanna Burger

Oil spills elicit images of oiled birds, struggling marine mammals, dead fish, large oil slicks, sludge, and tar balls washing ashore on pristine beaches and coastal environments. Massive oil spills, such as the *Exxon Valdez* in Prince William Sound off Alaska or the oil flowing into the Persian Gulf during the recent Persian Gulf War, remain before the public for many months as the media cover in detail the expansion of the spill, its effect on the environment, and the progress and efficacy of cleanup. These spills call into question the commitment and ability of industry practices and government regulations to prevent spills or to ensure adequate responses (Kelso and Brown 1991).

Media attention can be short-lived, however, and may be most prominent when the effects of the oil are most destructive. Interest dies down after a few weeks or months, particularly as the visual images become less powerful (Usher 1991). Massive media response focuses attention on environmental damage and often avoids covering cleanup or subsequent recovery (Mielke 1991). Media attention may also result in industry efforts to erase the fear of the general public toward oil spills and to minimize a spill's prominence in the media (Callis 1990). This sometimes results in oil industry officials standing amid oiled ecosystems and dying birds to explain that it is not all that bad.

The events in the Arthur Kill unfolded very differently, largely because the area was initially perceived as urbanized, polluted, degraded, and devoid of sensitive habitats (Swanson 1991; Burger, Parsons, and Gochfeld in press). Humans are intuitive toxicologists, using their senses of sight, taste, and smell to detect harmful or unsafe food, water, and air (Kraus, Malmfors, and Slovic in press), and during the Arthur Kill spill, the sights and smells were often overpowering. Yet between the industries and oil facilities are functioning salt marsh ecosystems with diverse trophic levels and even some endangered species (Burger, Parsons, and Gochfeld in press).

After the Exxon spill in the Arthur Kill media attention was slow to develop, until the key biological resources were described and defended by scientists and conservationists working in the Kill. Nonetheless, even throughout this spill the predominant image to reach the public was of people picking up oiled birds. The Arthur Kill, however, also encompasses a tidal salt marsh ecosystem with a variety of inverte-brates and vertebrates, not to mention the *Spartina* grass itself.

This chapter focuses on the immediate effects of the oil spill on se-lected biotic components of the ecosystem, "immediate" meaning the effects of the oil on the organisms living there when the spill occurred. Intermediate and longer-term effects on these same animals are covered more extensively in succeeding chapters. The selected organisms are *Spartina* grass, ribbed mussels (*Geukensia demissus*), fiddler crabs (*Uca* spp.), diamondback terrapins (*Malaclemys terrapin*), birds, and mam-mals. These are not the only organisms living in the Arthur Kill or affected by the oil; rather, they are used as indicators of how an oil spill can affect biota.

Timing of the Arthur Kill Oil Spill

Coastal ecosystems in temperate climates are heavily impacted by sea-sonal constraints such as changing temperature. The oil spill from the Exxon underwater pipeline was notable because it occurred in the win-ter. Because of this, many breeding birds were not in residence, winter-ing birds were present, some invertebrates and fish were under the mud or in underground burrows, migratory fish were out in oceanic waters, turtles were hibernating under the mud, and the *Spartina* grass was dormant. If the oil spill had occurred in the spring or summer the effects would have been more severe and perhaps even longer-lasting. The timing of the oil spill allowed the effects of an oil spill during a period of relative dormancy to be assessed.

Key Organisms Affected by the Spill

A discussion of the major components of the Arthur Kill ecosystem affected by the spill follows.

Cordgrass

Cordgrass, *Spartina alterniflora*, is the predominant plant fringing the creeks and channels (in natural salt marsh) ecosystems in eastern North America (Teal and Teal 1969). It grows in the peat from low tide to parts of the marsh reached by the very highest tides. Because it grows in the intertidal zone, it is at high risk from oil spills and other pollution events. Any toxic carried by tidal waters will eventually reach the *Spartina*, adhere to its stems, and ultimately affect its growth and survival.

The detrimental effects of oil on *Spartina* have been studied both in experimental studies and during natural oil spills (Baker 1971; Baker et al. 1989). Recovery often requires on the order of 5 to 10 years. This recovery period may be acceptable for ecosystems with a once-in-several-decades oil spill, but is more problematic for regions with frequent small spills.

The Exxon oil spill in the Arthur Kill occurred in conjunction with westerly winds that initially blew the bulk of the oil toward the Staten Island side of the Kill. Although the winds eventually shifted, the first wave of oil covered the *Spartina* along Staten Island, resulting in more severe damage there than elsewhere (Fig. 7.1).

Even though the *Spartina* was dormant in January and February, damage from the oil was noticeable, and the effect became more obvious when the new *Spartina* began to grow in May (Fig. 7.2). A bare fringe along the creeks and channels contained only the oil-stained, broken stems from the previous year's *Spartina*, while farther up the marsh the new green shoots were over a foot tall. By late summer 1990 a few scattered short, green shoots emerged through the oiled mud. In the summers of 1991 and 1992 more *Spartina* grew in this dead zone, but the effects were still marked and recovery slow (see Chapter 8 for more details).

Ribbed Mussels

Many invertebrates regularly live in the intertidal zone along the Arthur Kill, even though species diversity in the Kill is lower than in more pristine areas (Entrix and International Technolgy 1990). Ribbed mussels grow in profusion along the creeks and channels, jammed together vertically with their shells partially embedded in the mud. When covered with tidal waters, the shells open, and the animals within filter-feed. Ribbed mussels are common in salt marshes from Cape Cod to Florida, often concentrating near the high-water mark (Palmer 1949).

Normally in any bed of mussels very few are dead, although some shells remain long after predators, such as birds, have consumed the animal within. The distribution of dead mussels following the Exxon oil spill clearly indicated that oil contamination was very severe close to the spill, was intermediate at a few isolated locations and was less farther from the spill (Fig. 7.3). In some cases, such as Old Place Creek and adjacent creeks, mortality of ribbed mussels was almost 100 percent, and there were few live mussels in the summers of 1990 through 1993 (Burger unpub. data).

Fiddler Crabs

Whereas ribbed mussels represent invertebrates that are exposed to tidal action throughout the winter, fiddler crabs remain underground

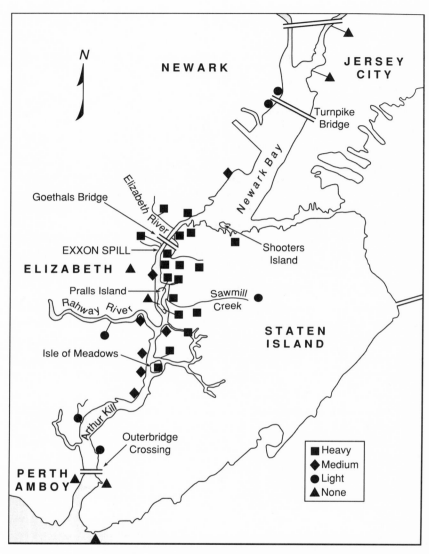

Figure 7.1. Effect of oil on *Spartina* and shoreline of the Arthur Kill (after Louis Berger & Associates, Inc., 1991).

Figure 7.2. View of *Spartina* in early February. For June see Fig. 4.1 (bottom).

from late fall until early spring. Fiddler crabs occur in eastern United States salt marshes from Cape Cod to Florida (Ward, Howes, and Ludwig 1976), providing an important food source for higher trophic levels such as birds. They are relatively sedentary, although they move about the mud flats near their burrows.

Previous data on the effects of oil on fiddler crabs were reported from West Falmouth, Massachusetts (Krebs and Burns 1977) and Nigeria (Snowden and Ekweozor 1987). Both spills occurred when the crabs were normally active, and both reported mortality. At Falmouth, fiddler crabs continued to show detrimental population effects for 7 years (Krebs and Burns 1977).

When the January 1990 oil spill occurred, fiddler crabs were 25 to 65 centimeters under the ground in vertical burrows, with a mud plug at the surface. Because of the oil spill fiddler crabs emerged prematurely from their underground burrows. Immediately after the spill, the oil covering the exposed tidal mud flats was mixed with dead and dying fiddler crabs of all sizes that were covered with oil (Fig. 7.4). Some crabs were relatively mobile, although they died within a day or two. Others were on their backs kicking their legs, and died within hours. The crabs were coated with oil.

Twenty-four creeks were censused in the Arthur Kill and Rahway

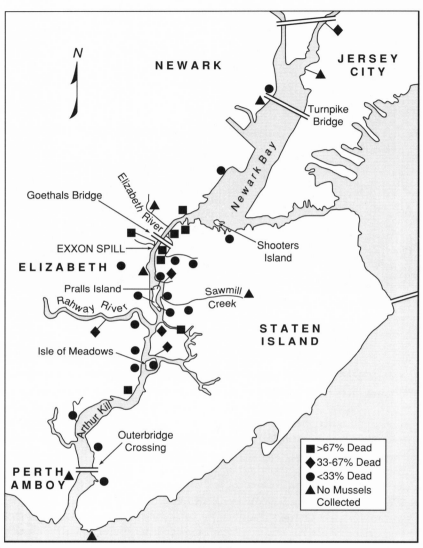

Figure 7.3. Percentage of dead mussels at different sample sites along the Arthur Kill (April–May 1990) (after Louis Berger & Associates, Inc., 1991).

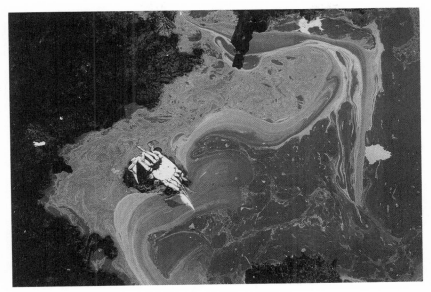

Figure 7.4. Dead fiddler crab on J's Creek, February 1990.

River, and six creeks were censused at nearby Cheesequake (28 km from the spill) to assess crab mortality (Burger, Brzorad, and Gochfeld 1992). Twice as many crabs emerged in the creeks adjacent to the spill compared with those that were over 4.5 kilometers (3 mi.) away. There were significant differences in the number of crabs that emerged as a function of distance from the spill (Fig. 7.5).

To assess the time course of the emergence, a 50-meter stretch along one creek across from the spill was surveyed at least once a week from immediately after the spill until April 3, 1990 (Burger, Brzorad, and Gochfeld 1992). During each survey all crabs were removed. Even so, on the survey from January 20 to March 14, over 50 percent of all crabs were dead, and the rest died within a few days in the laboratory. Initially, over a hundred crabs were collected each week, but by the end of January the number had dropped to just over seventy (Fig. 7.6).

That crabs continued to emerge for several weeks following the spill indicates the differential vulnerability of particular individuals. Crabs were forced to emerge when their burrows were soaked with oil. Nonetheless, variations in the physical location of the burrow (high or low on a mud bank), the depth of the burrow, the angle and width of the burrow, and the thickness of the mud plug no doubt determined the extent of oil contamination within their burrows and subsequent emergence. Some

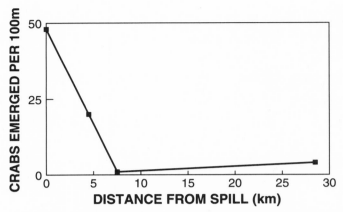

Figure 7.5. Number of crabs (per 100 m) emerging from sample creeks in the Arthur Kill and at Cheesequake (after Burger, Brzorad, and Gochfeld 1992).

perished in their burrows, and the number that died in this manner is incalculable.

Crabs that were alive upon collection were maintained in the laboratory so that their behavior could be observed (Burger, Brzorad, and Gochfeld 1991), their survival rates documented, and the effects of washing determined (Burger and Gochfeld 1992). It could be argued that once the fiddler crabs emerged they were doomed to die because of their premature emergence or because of their limited exposure to very cold air, rather than to the toxic effects of oil. To eliminate this possibility, fiddler crabs from the oil spill creeks were dug up from their underground burrows, kept in pails exposed to below-freezing temperatures while the mud was replaced, and subsequently maintained along with the oil-emerged crabs in the laboratory at room temperatures. Depending upon date of collection, only 29 to 45 percent of the oiled crabs survived for 2 weeks, whereas 97 to 100 percent of the dug-up crabs survived for 2 weeks. Only 2 percent of the oiled crabs survived until 5 weeks, and none made it to 8 weeks; whereas between 90 and 96 percent of the dug-up crabs were alive and healthy at 8 weeks, when the experiment was terminated.

Washing fiddler crabs with fresh water immediately after they came to the laboratory increased survival time and improved their ability to perform a variety of behavioral tests that involved balance, locomotion, and burrowing. Washed crabs showed great improvement in their ability to construct their own burrows. These experiments are important because they suggest that, for some invertebrates, allowing natural tidal action to bathe the mud flats may help to ameliorate the effects of oil.

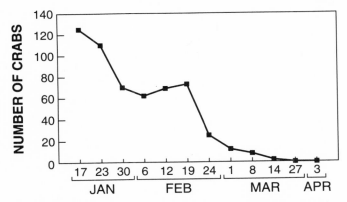

Figure 7.6. Emergence of fiddler crabs in one 50 meter stretch of creek directly across from the oil spill (after Burger, Brzorad, and Gochfeld 1992; Burger unpub. data).

Often booms placed to prevent oil from entering the creeks merely retain the oil that is already there. This happened in the Arthur Kill, because it was several hours before protective booms were placed at the entrance to creeks. In the meantime, more than a complete tide cycle and strong winds had ensured that oil was already in the creeks. Although the booms may have reduced new oil contamination, they also retained a heavy slick of oil in some creeks.

The behavior of oil-emerged crabs was also compared with that of crabs dug from the oil-contaminated creeks and with crabs that had already been maintained in the laboratory for several months (Burger, Brzorad, and Gochfeld 1991). Significant differences between the oil-contaminated and control crabs were apparent for all behaviors examined. Oiled crabs moved more slowly, took longer to right themselves when placed on their backs, could not grip on an incline, were less aggressive, and engaged in fewer defensive behaviors than either group of control crabs did (Fig. 7.7). All of these behaviors are essential for normal maintenance and survival in the wild. Crabs need to be able to defend themselves, escape quickly from aggressive conspecifics or predators, and move up and down muddy banks in search of food.

In total, these observations and experiments show that even in the winter, invertebrates that are buried in the mud are severely affected by an oil spill. All the crabs that emerged ultimately died, even when brought into warmer temperatures, and all showed behavioral impairments.

These were some of the first experiments conducted immediately after an oil spill with environmentally exposed (nonlaboratory) invertebrates.

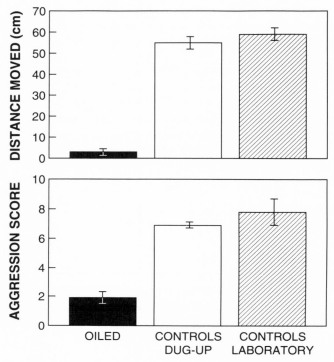

Figure 7.7. Comparison of oiled crabs that emerged following the oil spill with those dug up from the same area in January and some maintained in the lab for many months prior to the spill. Shown are the distances moved and aggression scores (mean ± SE).

Usually, as in the case of Falmouth (Krebs and Burns 1977), observations and experiments are initiated weeks or months after a spill, following government action. The lack of many observations within a month of a spill indicates the need for development of a funding mechanism to allow immediate scientific inquiry following a spill or other toxic event.

Diamondback Terrapins

Oil spills along coastal areas or in the open ocean are potentially dangerous for air-breathing organisms such as sea turtles (National Research Council 1990). Along the Atlantic and Gulf coasts of the United States diamondback terrapins are also vulnerable. They are small turtles (less than 25 cm long) that nest on sand dunes and barrier beaches but spend the rest of their lives in brackish water (Burger and Montevecchi 1975). During the winter they hibernate below the mud in the bottoms and sides of tidal creeks (Yearicks, Wood, and Johnson 1981). They remain dormant throughout the winter, emerging in the spring, when the mud temperatures rise. They are never above the surface in January.

Following the Exxon oil spill eleven diamondback terrapins were found covered with oil on the mud flats (Burger and Gochfeld in press). All eleven were female (normally only 60% of a population is female) [Yearicks, Wood, and Johnson 1981]), and all were weak. They were removed to the Staten Island Zoo for rehabilitation. Despite frequent changes of water, they continued to exude oil from their surfaces and to void oil from their digestive tracks 2 weeks after emergence. By a month after the spill all had developed edema, and their appetites were depressed. Only three survived to be released back to the Arthur Kill.

When tested 3 weeks after emergence the oil-exposed terrapins required 16 to 290 seconds to right themselves when placed on their backs, compared with an average of 2 to 3 seconds for terrapins examined in the summer, and 3 to 4 seconds for captive but nonoiled terrapins tested with the oiled terrapins. Similarly, oil-exposed terrapins scored 1 to 6 on a strength test, whereas winter captives and summer terrapins scored 8 to 10 (Burger and Gochfeld in press).

The effects of oil on diamondback terrapins, even in winter, are clearly severe. Those that died had oil traces in their tissues and exhibited digestive-tract lesions consistent with oil exposure. The differential mortality of females may reflect either differential vulnerability or that they hibernate closer to the surface, where oil can more easily penetrate to them. In either case, the end result is to potentially affect breeding populations, since females are more important than males to future production. A male terrapin can fertilize several females, but a female can produuce only so many clutches.

Birds and Mammals

Birds are the most visible animal component of many coastal ecosystems (O'Connor and Dewling 1986). They are large, numerous, and vulnerable to oiling, and the public responds quickly to a massive die-off of oiled birds. The visual image of an oil-covered cormorant, gull, or duck struggling to remain above the icy waves is very powerful.

In the aftermath of the Arthur Kill oil spill considerable media attention was focused on oiled birds, and a massive search, rescue, and rehabilitation effort was initiated. The rehabilitation efforts on wildlife, particularly birds, are discussed in Chapter 5. Therefore, I will only briefly summarize the results.

Oil affects birds directly by the oiling of plumage and indirectly

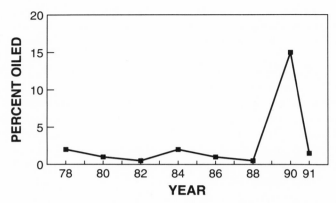

Figure 7.8. Incidence of oiled plumage in gulls in the Arthur Kill (after Burger, in press; Burger unpub. data).

through their food and water (Leighton 1982). Usually reports give the number of oiled birds found dead or moribund following an oil spill (e.g., Levy 1980). This approach, however, has two difficulties: It ignores a large majority of birds that have small exposures and do not die immediately, and it ignores the secondary effects such as transmission of oil to eggs during incubation (with lethal effects) (Lewis 1982) or to chicks (Boersma, Davies, and Reed 1988).

Search-and-rescue efforts in the Arthur Kill were conducted between January 3 and February 24, 1990 (see Chapter 5), when almost eight hundred birds were collected, over 80 percent of which were already dead. Almost 40 percent of the birds that were rescued alive were ultimately released back to the wild. Gulls made up 50 percent of the specimens collected or rescued, and another 41 percent were waterfowl. This reflects the timing of the spill. During the winter gulls and waterfowl use the Arthur Kill and Raritan Bay as an overwintering place (Burger, Parsons, and Gochfeld in press). Pathological examinations indicated that not all deaths were caused by the oil spill (see Chapter 5), underlining the importance of pathology to understanding the effects of oil spills on wildlife.

In addition to the birds, the search-and-rescue operations found twenty-eight muskrats (*Ondatra zibethicus*) that died during the oil spill. Although the Arthur Kill does not have any populations of marine mammals, land mammals that come to the Kill to forage or drink are also vulnerable to oil spills. The effects of oil on marine mammals that feed in coastal systems is discussed in Chapter 14.

One other indication of the severity of the Arthur Kill oil spill is the number of birds with oil on their feathers (Fig. 7.8). Immediately after the spill 15 percent of the gulls (*Larus*) had some oil on their feathers, although this dropped to 8 percent within 2 months. The drop could represent a washing off of oil or an exchange of gulls from elsewhere. Although in most cases the amount of oil was small, for some the oil spot was large enough to impair the thermoregulatory properties of the feathers.

Conclusions

Because several scientists were already working in the Arthur Kill prior to the spill, response was rapid and an assessment of damages was made immediately. These examinations of the immediate response of organisms to oil spills are essential to understanding the magnitude of the damage and to understanding the effects that became obvious only during the breeding season the following summer. This suggests that it is critical to mount a scientific inquiry within hours of the spill, to develop a mechanism to reimburse scientists and others that undertake these early studies, and to develop a mechanism to allow and facilitate access into the area by scientists.

Even though the oil spill in the Arthur Kill occurred in the dead of winter, devastating effects on the ecosystem occurred. These included damage to the *Spartina* grass, invertebrates such as fiddler crabs and ribbed mussels, and vertebrates such as diamondback terrapins, birds, and mammals. Only 20 percent of the *Spartina* was destroyed, but it was the 20 percent that fringes the creeks, channels, and Kill itself. This fringe along the tidal mud flats is where a large proportion of the algae and invertebrates live, and therefore where many fish, birds, and mammals feed.

That the oil had the effect of killing the fringing *Spartina* and intertidal mussels, as well as forcing fiddler crabs and diamondback terrapins from their overwintering places, is remarkable. It also suggests that many other invertebrates living in the soil were undoubtedly killed as well (see Chapter 9). The immediate death of a high proportion of the invertebrate populations in some areas inevitably results in lower population levels of these organisms in succeeding years, at least until population numbers can recover.

Decreases in organism numbers at these low trophic levels subsequently affects others, such as birds (see Chapter 13), that are higher on the food chain. This ripple effect can be felt for many years (Sanders et al. 1980). The length of time the effect can be detected depends not only on the magnitude of the original spill and hydrodynamics of the tidal estuary but on the occurrence of other spills. It is this latter factor that is particularly critical in the Arthur Kill.

The Arthur Kill area, including Newark Bay and the Kill van Kull, is highly industrialized, with numerous oil storage and transfer facilities. The large amount of traffic in oil tankers (see Chapter 1) increases the likelihood of a continuous series of oil spills. Although many will be small, others will be substantial (over 50,000 gallons). These repeated oil spills, like those that occurred in 1990, ensure that the ecosystem never has the time to recover. Thus, the system is perpetually in the "immediate effects" stage discussed in this chapter.

Only public opinion, public policy, industry actions, and government regulations can ensure that high-risk areas such as the Arthur Kill are protected from continual environmental insult. The ecosystem and its organisms possess resiliency and recuperative powers (see Chapters 8–13), but these can operate only if there is time for the ecosystem to recover without continued contamination.

Literature Cited

Baker, J. M. 1971. Seasonal effects of oil pollution on salt marsh vegetation. *Oikos* 22:106–110.

Baker, J. M., J. A. Bayleg, S. E. Howells, J. Oldham, and M. Wilson. 1989. Oil in wetlands. In B. Dicks, ed., *Ecological impacts of the oil industry*, pp. 37–59. New York: Wiley.

Boersma, P. D., E. M. Davies, and W. V. Reid. 1988. Weathered crude oil effects on chicks of fork-tailed storm petrel (*Oceanodroma furcata*). *Arch. Environ. Contam. Toxicol.* 17:527–531.

Berger, L., & Associates, Inc. 1991. Arthur Kill oil discharge study. Prepared for New Jersey Department of Environmental Protection.

Burger, J., J. Brzorad, and M. Gochfeld. 1991. Immediate effects of an oil spill on behavior of fiddler crabs (*Uca pugnax*). *Arch. Environ. Cont. Toxicol.* 20:404–409.

———. 1992. Effects of an oil spill on emergence and mortality in Fiddler Crabs *Uca pugnax*. *Environ. Monit. and Assess.*

———. 1992. Effects of washing fiddler crabs (*Uca pugnax*) following an oil spill. *Environ. Pollut.* 77:15–22.

Burger, J., and M. Gochfeld. In press. Diamondback terrapin (*Malaclemys terrapin*) affected by an oil spill when hibernating. *Environ. Pollut.*

Burger, J., and W. A. Montevecchi. 1975. Nest site selection in the terrapin *Malaclemys terrapin*. *Copeia* 1975:113–119.

Burger, J., K. Parsons, and M. Gochfeld. In press. Avian populations and environmental degradation in an urban river: The kills of New York and New Jersey. In J. A. Jackson, Jr., ed, *Avian Conservation*. Madison: University of Wisconsin Press.

Callis, C. F. 1990. Improving the public understanding of science. *Envir. Science and Technol.* 24:410–411.

Entrix and International Technology. 1990 (October). The environment and communities of the Arthur Kill. Report to Exxon Co.

Kelso, D. D., and M. D. Brown. 1991. Policy lessons from *Exxon Valdez* spill. *Forum for Applied Research and Public Policy* 6:13–19.

Kraus, N., T. Malmfors, and P. Slovic. In press. Intuitive toxicology: Expert and lay judgments of chemical risks. *Risk Analyses.*

Krebs, C. T., and K. A. Burns. 1977. Long-term effects of an oil spill on populations of the salt-marsh crab *Uca pugnax. Science* 197:484–487.

Leighton, F. A. 1982. The pathophysiology of petroleum oil toxicity in birds: A review. In *Proceedings of the Effects of Oil on Birds,* pp. 1–28. Wilmington, Del.: Tri-state Bird Rescue & Research.

Levy, E. M. 1980. Oil pollution and seabirds: Atlantic Canada 1976–1977 and some implications for northern environments. *Mar. Pollut. Bull.* 11:51–56.

Lewis, S. 1982. Effects of oil on avian productivity and population dynamics. Ph.D. diss., Cornell University, Ithaca, N.Y.

Mielke, J. E. 1991. Oil spills: Is the perception worse than the reality? *Forum for Applied Research and Public Policy* 6:5–12.

National Research Council. 1990. *Decline of the Sea Turtles.* Washington, D.C.: National Academy Press.

O'Connor, J. S., and R. T. Dewling. 1986. Indices of marine degradation: Their utility. *Environ. Manag.* 10:335–343.

Palmer, E. L. 1949. *Field Book of Natural History.* New York: McGraw-Hill.

Sanders, H. L., J. F. Grassle, G. R. Hampson, L. S. Morse, S. Garner-Price, and C. C. Jones. 1980. Anatomy of an oil spill: Long-term effects from the grounding of the barge *Florida* off West Falmouth, Massachusetts. *J. Mar. Research* 38:265–380.

Snowden, R. J., and I.K.E. Ekweozor. 1987. The impact of a minor oil spillage in the estuarine Niger delta. *Mar. Pollut. Bull.* 18:595–599.

Swanson, R. L. 1991. Ecosystem impact and use impairment in the New York bight. *Tidal Exchange* 2:1–2.

Teal, J., and M. Teal. 1969. *Life and Death of the Salt Marsh.* Boston, Mass.: Little, Brown.

Usher, D. 1991. Alaska, Gulf spills share similarities. *Forum for Applied Research and Public Policy* 6:27–29.

Ward, D. U., B. L. Howes, and D. F. Ludwig. 1976. Effects of temefos, an organophosphorous insecticide, on survival and escape behavior of the marsh fiddler crab *Uca pugnax. Oikos* 27:331–335.

Yearicks, E. F., R. C. Wood, and W. S. Johnson. 1981. Hibernation of the northern diamondback terrapin, *Malaclemys terrapin terrapin. Estuaries* 4:78–80.

8. Effects of Oil on Vegetation

Joanna Burger

When spills of oil or any other toxic chemicals occur, public attention is often directed at commercial losses of fish and shellfish, at property damage, at recreational losses, and at the death toll of conspicuous wildlife. For aquatic systems these concerns center around fish, marine mammals, and birds. This is understandable given their highly visible nature and their commercial value or importance for tourism. Nonetheless, one critical environmental need is for the public, scientists (both academic and industrial), conservationists, and government personnel to examine the effect of toxic chemicals on all aspects of the ecosystem. Too often the primary producers in an ecosystem, including the dominant vegetation, are ignored.

The vegetation in coastal habitats, particularly vulnerable to toxic chemical spills, is molded by a complex of environmental influences (tide, salinity, substrate movement) and biotic factors (dispersal biology, competition, succession, herbivory) (see Ehrenfeld 1990). Salt marshes fringe estuaries along most of the eastern Atlantic coast.

Often all successional stages, from seedlings on bare salt flats through mature high marsh vegetation supported by 8-meters-thick peat deposits, can occur within a short distance of each other. The development of a salt marsh depends upon the physiology of the local halophytic vegetation, the tidal regime, the processes of sedimentation, and changes in sea level (Redfield 1972).

Along the New Jersey coast *Spartina* spp. are the dominant salt marsh plants. Salt marshes fringe bays and estuaries and extend up rivers as far as salinity makes it possible for *Spartina* to outcompete freshwater vegetation. The short form of *Spartina alterniflora* grows in low-elevation marsh habitat with regular tidal inundation, and *Spartina patens* grows in the high marsh exposed to infrequent inundation yet higher soil salinity (Valiela, Teal, and Deuser 1978; Bertress 1991). *Spartina* salt marshes fringe the Arthur Kill wherever people have not filled or developed the land. Thus, it is important to examine the effect of the 1990 oil spill on *Spartina*, particularly because it is the base of the complex salt marsh ecosystem, ultimately supporting the invertebrates, fish, and birds that are such visible components.

By far the most extensive study of the effect of the 1990 oil spill on vegetation was conducted by International Technology Corporation (1990) and ENTRIX (Winfield and Vedagiri 1990) for Exxon in Linden,

New Jersey. Their reports serve as a basis for the analysis presented in this chapter. Initially, Exxon scientists planned to write this chapter but were constrained from proceeding. Their data on the effects of the spill, made part of the public domain by the settlement negotiations, are some of the most impressive and extensive.

Effects of Oil on Vegetation

Salt marshes are vital nurseries for the eggs, larvae, and juveniles of fish, crabs, shrimp, shellfish, and other species. Wetland areas are vulnerable to pollutants because oil and chemicals may become concentrated in these low-energy systems and persist for years (Bolze and Lee 1989). That is, most of a wetland area is not directly exposed to the cleansing surge of the surf, so only the edge that is exposed will experience the daily tidal inundation sufficient to wash away pollutants.

The most devastating effects of oil are from habitat losses or destruction, even if those losses are temporary, yet the effects on vegetation are often not mentioned in evaluations following oil spills and are seldom quantified. Usually only general statements are made concerning "habitat loss" (Maki 1991).

Oil can affect a variety of vegetation types. In tropical systems a major oil spill, such as occurred in Panama in 1986, can kill mangrove communities (Jackson et al. 1989). Five months after the Panama spill a band of dead or dying trees marked the zone where oil washed ashore, and by 6 months dead mangroves occurred along 27 kilometers of the coast (Jackson et al. 1989). Seedlings in the affected areas also died. Similarly, following the sinking of a barge in the Bonny Estuary of Nigeria nearly 30 percent of mangrove prop roots and 32 percent of seedlings were oiled (Snowden and Ekweozor 1987). Within a 500-meter area around the 250-barrel (=1,075 gallons) oil spill partial defoliation and death of the seedlings occurred.

The best data on the effects of oil on *Spartina* has been generated by J. M. Baker and her colleagues from England and Wales (Baker et al. 1990). In many estuaries *Spartina anglica* marshes coexist with oil industry installations and may be subject to a variety of oil-related disturbances (Baker et al. 1990). Crude oils and their products differ widely in their toxicity to *Spartina*. Toxicity increases along the series: alkanes (paraffins), cycloalkanes, olefins, to aromatics. Within a series, small molecules are more toxic than larger molecules (Baker 1970).

Experimental evidence shows that oil on *Spartina* leaves reduces the oxygen diffusion out of both the stems and roots; thus, oiling of aerial parts indirectly affects the underground system (Baker 1971a, 1971b).

Light oils penetrate plant tissue and disrupt membrane structure, while heavy oils smother plants and interfere with the normal oxygen diffusion process down the shoots and into the roots. A series of experiments involving different oil treatments to plots of *Spartina* indicated that it survives most single spillages by producing new growth from protected underground buds but does not tolerate chronic pollution (Baker 1976; Baker et al. 1980, 1984, 1990). Baker et al. (1990) note that *Spartina* can recolonize formerly oil-damaged areas.

Dispersant treatments are not effective in removing oil from *Spartina*, and cutting may increase damage, although it sometimes has been used as a cleanup treatment. Burning following an oil spill can result in new growth but is a difficult methodology in salt marshes. As an extreme measure *Spartina* can be removed, followed by seeding or transplanting, but this is a drastic and expensive cleanup technique.

Overall, the experiments of Baker and her colleagues indicate that die-offs of *Spartina* can occur following an oil spill. Under most circumstances, however, the *Spartina* can recover from an acute exposure more easily than from chronic exposure. Although these experiments are critical in elucidating how *Spartina* responds to oil of different types and under different exposure regimes, they do not quantify the actual effects of oil spills on natural salt marshes. The data from the Arthur Kill can be used to examine the effect of an oil spill of known magnitude on *Spartina* salt marshes.

The 1990 Oil Spill and Salt Marsh Vegetation

When the Exxon oil pipe burst in January 1990 it exposed 48 kilometers (30 mi.) of waterway and tributaries along the Arthur Kill to no. 2 fuel oil. Much of the undeveloped shoreline in the Arthur Kill, Kill van Kull, and Newark Bay area is dominated by cordgrass (*Spartina alterniflora*), although there are some salt hay (*S. patens*) patches. To even the casual observer, the *Spartina* along the Arthur Kill was covered with oil (see Fig. 8.1). Following the oil spill, Exxon contracted with International Technology Corporation of Edison, New Jersey, to conduct an injury assessment of *Spartina alterniflora*. This assessment (International Technology 1990) and the resultant analysis (Winfield and Vedagiri 1990) form the basis for the present examination. Their overall objective was to document the extent of the short-term injury to the intertidal salt marsh. Visual estimates of damage were compared with oiling level for the *Spartina* marshes in New Jersey and New York following the growing season.

Figure 8.1. View of *Spartina* marshes in the Arthur Kill following the 1990 oil spill.

Methods

Four wetlands ecologists assessed *Spartina* health in the field in the summer of 1990 when *Spartina* was green. They used transects that ran perpendicular to the shore. Transects were spaced 10 meters apart. To assess the condition of the *Spartina*, plots were divided into three zones (dead, recovering, and healthy). Dead *Spartina* was obvious as the stems were still oiled; black, not green, and obviously dead. Dead stalks ranged from 1 to 4 centimeters long and showed no evidence of new growth.

Recovering *Spartina*, often bordering dead *Spartina*, consisted of sparse new vegetative growth interspersed with the dead stalks remaining from the normal growth of the previous year. Some of the new growth had short underground rhizomes indicative of germination of seedlings (rather than recovery of existent *Spartina*). In other areas, recovering *Spartina* was taller, with mature rhizomes indicative of the previous year's growth. The below-ground growth consisted of dead and living rhizomes, indicative of sublethal effects of oil.

Healthy *Spartina* consisted of mature stands of *apparently* unaffected growth. Although these areas were classified as healthy, they did contain some remnants of oiled, dead stalks.

In most cases the transects were run for 33 meters from the lower

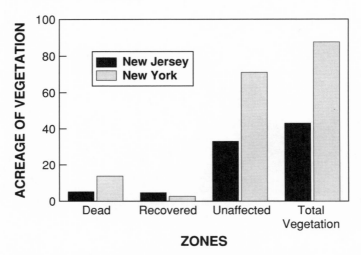

Figure 8.2. Acreage of *Spartina alterniflora* affected in New York and New Jersey by the 1990 oil spill.

limit of *Spartina* (at the peat edge) perpendicular to the shore. The width of the zones (dead, recovering, healthy) were recorded on field data sheets. Some transects were shorter than 33 meters because of intervening wrack lines, channels, or bulkheads. When a transect contained a dead or recovering zone wider than 33 meters, the transect was continued until a healthy zone was reached.

In addition, four levels of oiling were noted: heavily oiled, medium-oiled, minimally oiled, and no observed oiling. These categories reflect the amount of oil on the vegetation rather than the health of the *Spartina*. These categories were also noted at the end of the growing season.

Results

Some 48 kilometers of waterway were exposed to oil from the January 1990 Exxon spill. Four levels of oiling were subjectively delineated and three categories of damage noted. Approximately 52 hectares (131 a.) of *Spartina* were assessed in the studied transects, 35 hectares (88 a.) in New York and 17 hectares (43 a.) in New Jersey. In New Jersey about 2 hectares (5 a.) of *Spartina* were dead, and in New York about 5.5 hectares (14 a.) consisted of dead *Spartina* (Fig. 8.2). Despite the fact that initial media and scientific judgment was that New York was much harder hit than New Jersey, New Jersey lost a similar percentage (12%) of its marshes as New York did (15%). The recovering zones accounted for a

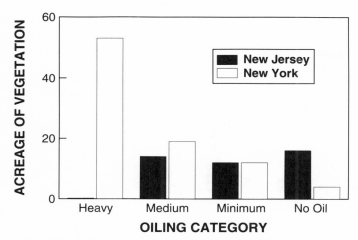

Figure 8.3. Degree of oiling of salt marshes in New York and New Jersey.

lower percent of the marshes than the dead zone (perhaps because some recovering vegetation was designated unaffected—see definitions previously given). Overall, of the 52 hectares (131 a.) of assessed *Spartina* marshes in both states, 7.6 hectares (19 a.) were dead, 2.8 hectares (7 a.) were recovering, and 42 hectares (105 a.) were unaffected. I feel compelled to note that the dead *Spartina* was, obviously, adjacent to the tidal mud flats, the interface that is the zone of highest use by invertebrates, fish, and birds (see the discussion that follows). Thus, the ecological impact was disproportionate to the actual spatial extent.

The categorization of oiling (Fig. 8.3) indicates that the heaviest oiling occurred along the New York marshes. Less than 2.5 hectares (1 a.) of *Spartina* in New Jersey was categorized as experiencing heavy oiling. Most *Spartina* in both states experienced some oiling; only 8 hectares (20 a.) had none.

From a biological viewpoint, the width of the impacted *Spartina* zone in relation to the high-tide line is critical because this is the most highly used foraging zone by invertebrates, fish, and birds. The average width of the dead *Spartina* zone in New York was nearly 3 meters compared with 1.5 meters for New Jersey (Fig. 8.4). Averaged across the entire shoreline, it is remarkable that only 3 meters was affected. When the recovering zone is added, the affected *Spartina* zone is even wider.

It is instructive to compare the degree of oiling with the effect categories (Fig. 8.5). This assumes consistency in each of these measures. That is, "heavy" is always categorized the same way, and "dead" is always categorized the same way. As expected, the amount of unaffected

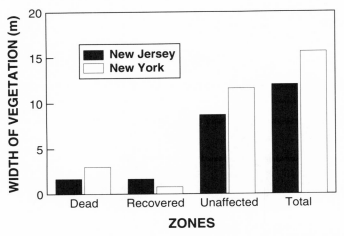

Figure 8.4. Average width of *Spartina* in New York and New Jersey affected by the oil spill.

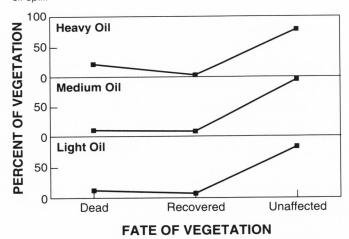

Figure 8.5. Relationship between the oiling categories and the effects classes (dead, recovered, unaffected *Spartina*) following the oil spill.

Spartina increases with decreased oiling: When there is less oil, less *Spartina* is affected.

A similar relationship exists for the oiling categories and the width of affected zones (Fig. 8.6). The *Spartina* zone was widest in areas of heavy oiling and narrower with minimum oiling. Interestingly, a small band of dead *Spartina* appeared in the no-oil category, indicating that the oil was no longer visible but had had an effect.

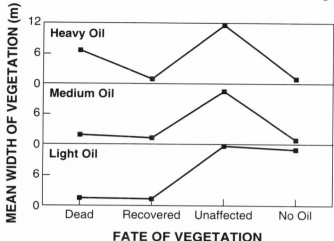

Figure 8.6. Relationship between the oiling categories and the width of damage to *Spartina* following the oil spill.

In 1993, three growing seasons after the 1990 oil spill, most of the dead *Spartina* marsh had still not recovered. In some areas originally designated by Winfield and Vedagiri (1990) as "recovering," however, the *Spartina* was lusher and taller than in 1990. Similarly, a few of the areas designated as dead in 1990 showed some signs of recovery. A few short (less than .5 m tall) *Spartina* stems were invading these dead zones via underground rhizomes from the nearby recovering zones. Other parts of the dead *Spartina* marsh fringing the Arthur Kill or salt marsh creeks and channels remained dead, the dead stems long since decayed and washed away by high tides. In years to come the *Spartina* will continue to recover, given the proximity to healthy *Spartina* and the absence of any additional major oil spills. However, the effects on pro-ductivity and on other populations will persist for several years.

Discussion

Overall, the results of this study indicate that New York had nearly twice as many *Spartina* marshes in the affected area as New Jersey did. Thus, the greatest extent of visual injury was in New York, but the relative proportion of dead *Spartina* in each state was similar. In the growing season following the oil spill, 15 percent of the acreage was dead and 6 percent was recovering. However, it was the 21 percent adjacent to the tidal mud flats and creeks that was affected. The average width of the affected *Spartina* was greater in New York than in New Jersey.

Three conclusions are of general interest: (1) A substantial proportion of the *Spartina* was affected, (2) the affected areas were adjacent to tidal mud flats, and (3) New York and New Jersey were equally affected, even though the initial damage was more apparent on the New York side. I discuss each of these briefly.

Overall, only 21 percent of the total *Spartina* examined was affected by the 1990 Exxon oil spill; nearly 80 percent was unaffected. The importance of this fact depends to some extent upon one's perspective. To lose over 20 percent of any productive area by only one event has serious consequences, particularly in an ecosystem that is also exposed to chronic oil spills and that may be exposed to other major oil spills before the area is completely recovered.

There are two problems with the methodology of this study: (1) The estimation of damage to *Spartina* was made on the basis of visual estimates, and (2), even the "unaffected" area had some dead stems apparently killed by the oil. Thus, the actual damage no doubt affected more than the estimated 21 percent.

Furthermore, oil seeps into the soil, remaining for many years. Oil that penetrated the salt marshes following an oil spill in the West Falmouth marshes remained even 20 years later (Sanders et al. 1980). To adequately examine damage to *Spartina* it would be necessary to compare growth pattern (height, stem thickness, seed viability) and biomass production of above- and below-ground vegetation in areas that were, and were not, exposed to the oil. That is, *Spartina* vegetation may appear green and healthy in the impacted area yet still be less than healthy (and thus affected).

The areas of *Spartina* affected were adjacent to the tidal interface. Since oil moved into the marsh via the tidal creeks and channels as well as via the Arthur Kill itself, the greatest oil deposition occurred along the shore. Less oil penetrated the high marsh, particularly since the tides were not excessively high immediately following the spill.

To many people, the fact that nearly 80 percent of the *Spartina* was apparently unaffected was reassuring and indicative of the level of damage. However, the interface of salt marsh vegetation with tidal waters is one of the most productive areas of the marsh for invertebrates and vertebrates. A number of species, such as mussels (see Chapter 9), live exclusively in this zone; while others, such as fiddler crabs (see Chapter 10), are most abundant there. Thus, any damage to this zone, even though it is a small percentage of the total ecosystem, has a far greater effect than mere percentages suggest.

The salt marshes of the Arthur Kill are one system, regardless of whether the marshes are in the jurisdiction of New York or New Jersey. The mapping done following the 1990 oil spill indicated that more marshes exist on the New York side than on the New Jersey side of the Arthur Kill. New York had more potential losses from the spill. The data examined in this chapter, however, indicate that both states suffered similar losses of *Spartina* in terms of percentages. Thus, although the initial damage appeared far more severe on the New York side following the growing season, the rate of damage was similar. Partially this relates to the subsequent movement and distribution of the oil by normal tides in the months following the spill.

Conclusions

Spartina is the primary vegetation in temperate salt marsh or estuarine habitats and is vulnerable to major oil spills. In preparation for an oil spill, or any type of environmental degradation (other toxic event or even development), communities should have wetlands maps prepared that generally designate the extent of *Spartina alterniflora* and *S. patens* as well as noting the margin of incursion of upland plant species. To some degree, knowing the extent of *Spartina alterniflora* will give some indication of the potential damage exposure of the marsh by acute exposure to oil or other toxic because this is the zone of daily tidal inundation as well as the zone most likely to be impacted by chronic exposure.

For most animal resources it is essential to have some indication not only of presence but of abundance. This is not as true for vegetation, however, because sampling of oiled and unoiled salt marsh plots following a spill can be used to document and assess damages. In the Arthur Kill this assessment took the form of a visual evaluation (dead, recovering, unaffected), which was sufficient to estimate the immediate effects. This evaluation, however, is not sufficient for a temporal evaluation over many years, because the long-term effects, particularly for the recovering and unaffected zones, will relate to overall productivity. That is, the presence of oil in the salt marsh sediments may continue to result in decreased productivity (less dense growth, shorter plants) for several years, even though a given patch may appear healthy to the casual observer.

Although Baker and her colleagues (see "Effects of Oil on Vegetation," above) have examined the effects of oil on *Spartina* experimentally, the long-term effects of an oil spill on *Spartina* in nature have not been measured. Since *Spartina* is, in most cases, the base of the salt marsh ecosystems, any decrease in productivity will have a domino effect on the whole system.

For most of the biota discussed in this book, documentation of presence and abundance is the extent of community preparedness recommended. For *Spartina* vegetation, however, the presence of an oil spill plan and ready equipment for implementation of the plan may have an important effect on the damage sustained. That is, since the fringe of salt marshes is exposed to tidal water, exposure can be limited by limiting tidal access.

Tidal access obviously cannot easily be controlled for many estuaries or large rivers, but it can be controlled for tributaries such as small channels and creeks. Immediate placement of absorbent booms to prevent oil from entering these small tributaries can have a major effect on the amount of oil that reaches the interior of marshes. It is critical, however, to place these booms immediately, before oil has entered the creeks, otherwise the booms will only keep oil in the marsh, enhancing its impact rather than allowing tidal flushing to remove it.

Community preparedness for an oil spill or any environmental insult should involve an overall mapping of wetland resources, the ability to initiate a detailed study of effects, and a plan for immediate response. In all three cases, prior identification of scientists, appropriate government officials, and oil response personnel is critical for success.

Acknowledgments

I thank the personnel of Exxon and the New Jersey Department of Environmental Protection and Energy for so kindly providing me with reports and insights concerning the Exxon studies on the vegetation of the Arthur Kill, and J. Baker for valuable discussions and reprints on the effects of oil on vegetation. I also thank John Brzorad and Michael Gochfeld for comments on the manuscript.

Literature Cited

Baker, J. M. 1970. The effects of oil on plants. *Environ. Poll.* 1:27–44.
———. 1971a. The effects of oil pollution and cleaning on the ecology of salt marshes. Ph.D. diss., University of Wales.
———. 1971b. Studies on saltmarsh communities. In E. B. Cowell, ed, *The Ecological Effects of Oil Pollution on Littoral Communities*, pp. 16–101. Barking, England: Applied Science.
———. 1976. Ecological changes in Milford Haven during its history as an oil port. In J. M. Baker, ed., *Marine Ecology and Oil Pollution*, pp. 55–66. Barking, England: Applied Science.
Baker, J. M., J. H. Crothers, D. I. Little, J. H. Oldham, and C. M. Wilson.

1984. Comparison of the fate and ecological effects of dispersed and non-dispersed oil in a variety of intertidal habitats. In T. E. Allen, ed., *Oil Spill Chemical Dispersants: Research, Experience and Recommendations*, pp. 239–279. Philadelphia, Pa.: American Society for Testing and Materials.

Baker, J. M., J. H. Crothers, J.A.J. Mullett, and C. M. Wilson. 1980. Ecological effects of dispersed and non-dispersed crude oil: A progress report. In *Proceedings of the Institute of Petroleum Conference on Petroleum Development and the Environment*, pp. 85–100. Heyden, London.

Baker, J. M., J. H. Oldham, C. M. Wilson, B. Dicks, D. I. Little, and D. Levell. 1990. *Spartina anglica* and oil: Spill and effluent effects, clean-up and rehabilitation. In A. J. Gray and P.E.M. Banham, eds., Spartina anglica: *A Research Review*, pp. 52–62. London: HMSO.

Bertress, M. D. 1991. Zonation of *Spartina patens* and *Spartina alterniflora* in a New England salt marsh. *Ecology* 72:138–148.

Boersma, P. D., E. M. Davies, and W. V. Reid. 1988. Weathered crude oil effects on chicks of fork-tailed petrels (*Oceanodroma furcata*). *Arch. Environ. Contam. Toxicol.* 17:527–531.

Bolze, D., and M. Lee. 1989. Offshore oil and gas development: The ecological effects beyond offshore platforms. *Proceedings of the Sixth Symposium on Coastal and Ocean Management:* 1920–1934.

Ehrenfeld, J. 1990. Dynamics and processes of barren island vegetation. *Rev. Aquatic Sciences* 2:437–480.

International Technology Corporation. 1990. *Spartina alterniflora* injury assessment study. Report to Exxon Co.

Jackson, J.B.C., J. D. Cubit, B. D. Keller, V. Batista, K. Burns, H. M. Caffey, R. I. Caldwell, S. D. Garrity, C. D. Getter, C. Gonzalez, H. M. Guzman, K. W. Kaufman, A. H. Knap, S. C. Vevings, M. J. Marshall, R. Steger, R. C. Thompson, and E. Weil. 1989. Ecological effects of a major oil spill on Panamanian coastal marine communities. *Science* 243:37–44.

Maki, A. W. 1991. The *Exxon Valdez* oil spill: Initial environmental impact assessment. *Environ. Sci. Technol.* 25:24–29.

Redfield, A. C. 1972. Development of a New England salt marsh. *Ecol. Managr.* 42:201–237.

Sanders, H. L., J. F. Grassle, G. R. Hampson, L. S. Morse, S. Garner-Price, and C. C. Jones. 1980. Anatomy of an oil spill: Long term effects from the grounding of the barge *Florida* off West Falmouth, Massachusetts. *J. Marine Research* 38:265–280.

Snowden, R. J., and I.K.E. Ekweozor. 1987. The impact of a minor oil spillage in the estuarine Niger delta. *Marine Pollut. Bull.* 18:595–599.

Valiela, I., J. M. Teal, and W. G. Deuser. 1978. The nature of growth forms in the salt marsh grass *Spartina alterniflora*. *Amer. Natur.* 112:461–470.

Winfield, T., and U. Vedagiri. 1990. Visual effects of a no. 2 fuel oil spill on the intertidal *Spartina* marshes along the Arthur Kill and tributaries. Report to Exxon Co.

9. The Effects of Oil Spills on Bivalve Mollusks and Blue Crabs

Keith R. Cooper and Angela Cristini

The presence of petroleum-derived hydrocarbons in estuaries and near-shore coastal waters is a consequence of life in an industrial society. Chronic inputs as well as oil leaks such as the 1990 Exxon spill in the Arthur Kill add to the burden of organic compounds present in the waters and sediments of the Arthur Kill. The effects described in this chapter regard studies on no. 2 fuel oil that are found in the literature and observed effects on the fauna and flora inhabiting a small *Spartina* marsh and mud flat located in Elizabeth, New Jersey (Fig. 9.1). We have been monitoring invertebrate and lower vertebrate species at this marsh since 1985. This site was heavily oiled during the 1990 spill due to the placement of absorbent booms just north of the mud flat, resulting in the collection of oil along the shore.

In the first few days following the spill, ribbed mussels (*Geukensia demissus*), soft-shelled clams (*Mya arenaria*), and the Balthic macoma clam (*Macoma balthica*) suffered heavy mortalities at the Elizabeth site. Although not quantified, many clams were found gaping and dead in the sediments. The oil penetrated into the mud flat, leaving other invertebrates dead as well. These include the mud snail (*Melampus bidentatus*), periwinkles (*Litorina litorina*), and the blood worm (*Glycera americana*). The mud flat was devoid of any bivalves or other invertebrates until the late spring and early summer of 1990, when recolonization began. On the upper portion of the mud flat were rocks and peat islands with *Spartina* grass. The peat was soaked with oil, and even through the summer of 1991 oil bubbled out of the peat when the peat was walked on. Pools of oil had collected underneath rocks and persisted for months following the spill. The ribbed mussels that lived in the peat were killed, and the cordgrass (*Spartina alterniflora*) growth on the peat was reduced throughout the entire first summer. On heavily oiled peat no growth was observed. Hampson and Moul (1977) reported that marsh grasses in a small cove oiled by no. 2 fuel oil in Massachusetts were unable to reestablish themselves 3 years after the spill. Similar observations were made at the Elizabeth, New Jersey, site following the 1990 oil spill. Oil was trapped in the rocks, and soft-shelled clams living in these areas continued to show mortality throughout the winter months of 1990. Because of the shell morphology of soft-shelled clams, they cannot completely close their shells, and their mantle and gill surfaces are always exposed to sediment and interstitial water.

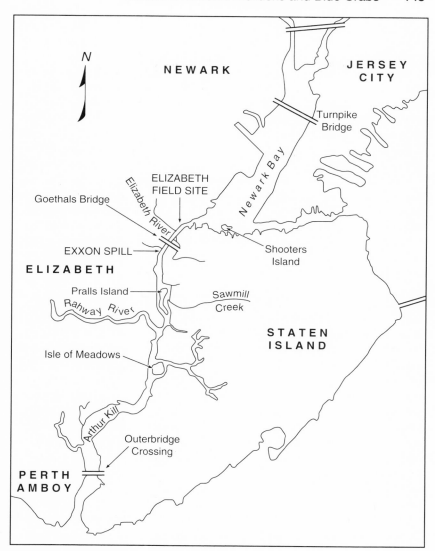

Figure 9.1. Location of Elizabeth field site.

From February until September 1990 an estimated 2,000,000 liters of petroleum products spilled into the Arthur Kill. None of these subsequent spills appeared to reach the Elizabeth study site. In an urban community such as the New York–New Jersey harbor estuary a number of pollution sources other than oil impact the communities living in these waters (Squibb et al. 1991). Even though this is an urban estuary it supports a wide variety of aquatic animals and serves as a major nursery area for both aquatic and avian species (see Chapter 1).

One of the important members of the benthic community of the Arthur Kill is the blue crab (*Callinectes sapidus*). Our observations indicate that large populations of this species inhabit the Arthur Kill as well as Newark Bay. Juveniles and adults live on the sediments, consume a variety of benthic invertebrates, and are an important source of food for demersal fish. The bivalve communities are also an important food source for other species in this area such as migratory birds and finfish. The soft-shelled clam is one of a number of species that live buried in the mud flats and sediments of the Arthur Kill. Because of the importance of these two species to the benthic ecosystem, this chapter highlights the effects of petroleum-derived hydrocarbons on them. The conclusions drawn from these two species can be generalized to other species within their classes.

Effects of Oil on Invertebrates

The type of oil spilled is very important for determining the oil's effects on aquatic animals as well as for influencing the duration of its impact. The environmental conditions and time of year when the spill occurred also alter the biological ramifications. Even though generalizations can be made about refined versus crude oils, the toxicity within each type of oil varies (Anderson 1975). Anderson (1975) demonstrated that in crustaceans and finfish the refined oils, both as water-soluble fractions and as oil–water dispersions, were significantly more toxic than the nonrefined crude oils. The toxicity of oil types is also species dependent, but the LC50 (the dose that will kill 50% of the organisms) for the water-soluble fraction (WSF) for no. 2 fuel oil was 3 to 4 parts per million for crustaceans, while for crude oil it was 8 to >20 parts per million (Anderson 1975). The majority of the toxicity associated with the WSF was attributed to the naphthalenes and the substituted methylnaphthalenes. These compounds are water soluble and accumulate in the tissues of organisms.

Acute effects occur over a short period of time following an oil spill. Deaths during this early time period are normally due to very high dosages of water-soluble components, physical clogging, and morphological damage to the animal's respiratory surfaces. One of the reasons that these effects occur shortly after a spill is that the oil spreads on water and dissipates over larger areas with time. The more volatile and toxic components are lost either into the air or diluted into the water column. If the oil becomes trapped in the sediments or vegetation then the acute toxic effects will be observed for much longer periods of time. This is due to the reduced diluting effects of the water and air and the decreased availability of bacterial degradation.

Animals and plants that were present in the waterway suffered extensive losses during the 1990 Exxon spill (see Chapter 7). Fortunately, even an urban estuary is dynamic and can recover from an oil spill when populations either surviving the spill or living outside the impacted area can repopulate disturbed areas. The persistence of oil in the environment and within organisms is limited if the source of contamination is eliminated. Mobile invertebrates and fish that survive the acute exposure can move to less contaminated areas after a spill. Most bivalve mollusks, however, are sedentary and are unable to move to cleaner areas, but in some cases they can close their shells to prevent exposure. Eventually, though, bivalve mollusks must open their shells to respire and feed.

Bivalve mollusks exposed to oil have been reported to have depressed gill tissue oxygen consumption and decreased filtration rates and will shift to anaerobiosis (Anderson 1973; Thurberg, Gould, and Dawson 1978). Stainken (1978) reported the effects of no. 2 fuel oil on the respiratory rates of *Mya arenaria* and correlated them with tissue levels of petroleum-derived hydrocarbons. These studies are relevant to the situation that occurred following the 1990 Exxon spill since they were carried out at 4°C at concentrations of 10, 50, and 100 parts per million. At low concentrations in *Mya* the respiratory rate doubled, while at higher concentrations a decrease in the respiratory rate was reported. Galtstoff et al. (1935) reported that the WSF from crude oil resulted in anesthetic effects on the ciliated epithelium of gills. At high concentrations of oil animals reduce filtration rates and shift the energy budget to a more anaerobic situation. Gilfillan (1973, 1975) reported alterations in the carbon budgets of soft-shelled clams exposed to crude oil. This same author reported that after 3 years following a no. 6 fuel oil spill, and 6 years after a crude-oil spill, soft-shelled clams had lower growth, reduced carbon flux, and a lowered assimilation rate (Gilfillan and Vandermeulen 1978).

An additional response of bivalves to irritants in general is increased mucus production to protect the epithelium, which could be damaged by these compounds. The production of these mucopolysaccharides results in the animal expending additional energy for the synthesis of these materials, creating more stress for the animal. The production of these mucopolysaccharides also decreases the gills' efficiency to transport oxygen across the gill.

Effects on Invertebrates Following the 1990 Exxon Oil Spill

Studies on the soft-shelled clam carried out following the Exxon spill gave very similar findings to those reported by other researchers. We had been sampling soft-shelled clams from the Elizabeth, New Jersey, site since 1985, allowing the establishment of a baseline level of lesions prior to the Exxon spill. A lesion is defined as a morphological change from normal tissue structure. Lesions can be caused by parasite infestation, trauma, or chemical insult. The alteration from normal can vary from slight to extensive change and is ranked according to severity. Because of the repair capability of animals, damaged tissue can return to normal. This is like the repair of a cut in humans, but if the cut is extensive a scar will remain. Similar repair occurs in invertebrates.

Comparisons of oiled clams were made with baseline lesion occurrence established in 1988 and 1989. Approximately thirty animals were collected 1 week following the spiill and at monthly intervals thereafter through December 1990. The animals were fixed in 10 percent buffered formalin and evaluated by light microscopy. All of the major organs and tissues were evaluated for histopathological lesions. Following the system used by Berthou et al. (1987), the lesions were graded as absent, mild, moderate, or severe. Both moderate and severe lesions result in impairment of normal function. The gills showed focal hyperplasia, and hemocytic accumulation was found in the gills. Debris and dark-stained material also accumulated. In several of the animals the gill lamellae were fused. This is similar to reports on Atlantic bay scallops (*Aequipecten irradians*) and American oysters (*Crassostrea virginica*) exposed to waste motor oil and fuel oil (Gardner, Yevich, and Rogerson 1975). The incidence of lesions (total) in the gills following the spill were not significantly different from prespill clams. This is due to the fact that these animals are not living in a pristine environment, and the gills respond to a number of toxic chemicals in a similar manner. The gill lesions in the clams from Elizabeth are typical of soft-shelled clams living in a chronically impacted environment.

The results from these studies are shown in Table 9.1 and Figure 9.2. The incidence of moderate to severe lesions of the digestive gland, intestine, and mantle increased rapidly from mid 1988 to the end of 1990. The digestive gland showed hypertrophy, hyperplasia, metaplasia, and increased basophilic staining of the epithelial cells compared with prespill clams (chi-square $p < .05$). These lesions in the digestive gland, intestine, and mantle were due to the oil spill.

The mantle also showed hyperplasia, vacuolization, and mucous-cell proliferation (chi-square $p < .05$). The lesions that were observed were

Table 9.1. Incidence of *Mya arenaria* with Moderate to Severe Lesions

Collection		Tissues					
Year	Month	Digestive Gland	Gill	Mantle	Intestine	Kidney	Heart
1988	July	—	30	—	11	25	0
	September	32	41	47	17	8	0
	December	35	14	18	17	67	0
1989	February	30	4	28	18	67	8
	March	—	—	—	—	—	0
	April	33	24	33	19	60	0
	May	—	45	—	13	50	0
	July	—	6	0	0	7	0
				Rupture of pipeline			
1990	January	88[a]	40	69[a]	77[a]	60	0
	February	45	4	31	52[a]	48	0
	March	33	22	20	25	86	8
	April	44	32	28	57[a]	95[a]	20
	May	6	23	23	32	38	—
	June	27	8	11	23	36	0
	July	39	50	21	29	59	0
	September	54	17	ND	21	58	0
	November	4	12	9	7	27	12
	December	17	11	ND	4	37	0

SOURCE: Brown, Cristini and Cooper (1992).
NOTES: Collected from Elizabeth, New Jersey.
Values are percentage response N = 30. — = not examined. ND = nondetectable.
[a]Chi-square p < .05 comparison from prespill percentages.

similar in a number of ways to those reported following chemical wound healing reported by Sparks and Pauley (1965) in the Pacific oyster (*Ostrea lurida*) and observed in other mollusks (Sparks 1972).

The intestinal lesions occurred very rapidly and persisted for about 4 months after the spill (chi-square p < .05). A delayed effect was observed in the kidney and heart, which may have been due to other spills of crude oils in the area. Stainken (1976) reported a depletion of glycogen and a generalized hemocytosis in a number of blood sinuses along the pallium and mantle. He also reported increased vacuolization of the gastrointestinal tract. Following the *Amoco Cadiz* spill in France both flat oysters (*Ostrea edulis*) and giant Pacific oysters (*Crassostrea gigas*) were monitored for histopathological effects for 7 years (Berthou et. al. 1987). The digestive tract had the highest incidence of necrosis, followed by interstitial tissue and gills. The interstitial-tissue lesions could have been due to parasitism. Severe gonadal alterations were observed in the first 2 years for the flat oysters, but no similar effect could be shown for

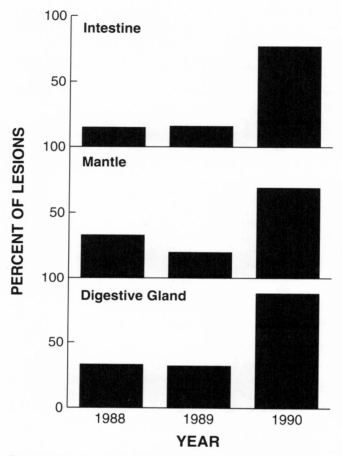

Figure 9.2. Percentage of lesions on the mantle, digestive gland, and gill of *Mya arenaria* clam in 1988 and 1989 (before the spill) and in 1990.

the giant Pacific oysters (Berthou et al. 1987). These authors also indicated the difficulty of establishing clear causes of histological lesions except during high exposure concentrations.

The temporal occurrence of the lesions and organ systems involved in the Arthur Kill animals were similar to those reported from the *Amoco Cadiz* spill. The duration of the effects was longer following the *Amoco Cadiz* study and may reflect the fact that the animals living in the Arthur Kill are exposed to spill situations chronically, while the oysters in France had never been exposed.

Several reports indicate that invertebrate and lower-vertebrate species that are resident in the Arthur Kill are more resistant to exposure from contaminants than the same species from pristine sites; killifish (*Fundulus heteroclitus*) (mercury and dioxins), soft-shelled clams (dioxins and transplantation effects on Adenylate energy charge) (Weis et al. 1986; Cristini 1987; Prince 1990; Brown 1991).

The gonadal development of soft-shelled clams in the Arthur Kill was not affected by the spill, as both male and female development appeared similar to that of nonimpacted populations of soft-shelled clams. Similarly, the siphon, gill, stomach, foot, muscle, and body epithelium were not found to be adversely affected by the oil spill.

The bioconcentration factor for oil has been estimated for several different bivalves to be between 2,000 and 4,000. It must be realized that oil is a complex mixture and that this represents the averaging of several compounds (Stegeman and Teal, 1973, Anderson 1975; Fossato and Canzonier 1976; Berthou et al. 1987). As with most xenobiotics the uptake is determined by passive diffusion either across the gills, epidermis, or mantle, or through the gastrointestinal tract. The saturated hydrocarbons appear to be rapidly lost, while polyaromatic hydrocarbons are more slowly released or are maintained. No information exists that demonstrates food web magnification in fish or higher vertebrates because of their ability to metabolize the aromatic compounds.

There is evidence that benzo(a)pyrene and its conjugates can be concentrated in lower invertebrates and that at higher trophic levels it can accumulate in the green gland of lobsters (James et al. 1991). There have been reports of mixed-function oxidase activity in a number of mollusks and crustaceans and that the activity is generally highest in the digestive gland or hepatopancreas (Singer and Lee 1975; James and Shiverick 1984; Anderson 1985; Livingstone 1985). There is also evidence of conjugation reactions that aid in the elimination of lipophilic compounds.

Elimination of aliphatic hydrocarbons occurs very rapidly, in the order of 3.5 days for the blue mussel (Fossato 1974). The elimination of oil hydrocarbons appears to follow at least a biphasic pattern with a very rapid phase with the time needed to decrease tissue levels by 50 percent of 3.5 days and then a slow phase of incomplete elimination with 12 percent remaining at the end of 8 weeks after being placed into clean water. These studies were important because the mussels had been exposed to the hydrocarbons for a long period of time and were at equilibrum upon moving to clean water. These authors also reported that animals with lower levels of hydrocarbon loads did reach background levels within 20 to 40 days. Similar findings were reported from Pacific and flat oysters (Berthou et al. 1987). The elimination of the hydrocarbons was independent of temperature as long as the animals were filtering water. When the filtration rate dropped due to temperature, however, all elimination also stopped. These studies are in agreement with a number of earlier and later studies that indicate that elimination appears to be biphasic, with a rapid phase followed by a more prolonged second phase.

The explanation for the second phase is that compounds are sequestered in compartments that are not well perfused by blood and therefore are not readily able to release the compound into circulation so that it can be eliminated. These compartments also contain lipids, with which the lipophilic (lipid-loving) compounds tend to associate. Since the movement of these compounds is by passive diffusion they prefer to remain within the tissue and move slowly into the less lipid-rich blood. The blood carries these compounds to the site of elimination (kidney, gills, and mantle). The shorter the exposure period, the faster the rate of elimination (Berthou et al. 1987). This type of biphasic elimination has been modeled for the bivalve *Mya arenaria* and can be explained using a pharmacokinetic-based model. The explanation for the relatively fast rate of loss is due in large part to the tremendous volume of water that bivalves can filter. The passive diffusion is concentration gradient driven, so that when the water passing over the surfaces of the animal contains low concentrations of contaminants, the net movement will be toward the water. Tissues that are poorly vascularized will take longer to rid themselves of the contaminants.

Behavioral effects on the fiddler crab following the Exxon spill into the Arthur Kill were reported by Burger, Brzorad, and Gochfeld (1991) and are discussed in Chapters 7 and 10. The behavioral effects reported are not unexpected since a number of the compounds present in freshly spilled no. 2 fuel oil cause a narcosis-type (anaesthetic) response. This is manifested as locomotor-type effects. The systemic toxicity could result in death due to direct organ damage (respiratory damage).

In the case of the blue crabs inhabiting this impacted area, similar effects would be expected if the oil reached the crabs buried in the sediments. The blue crab could then ingest the oil, or high levels of oil could result in the death of the blue crab and any blue crab predators (Brown and Cooper 1978). Oil in the water column has been shown to accumulate on zooplankton, which can also serve as a means for exposure of oil to zooplankton predators (Polak et al. 1978). The time of the spill (in winter) limited the amount of zooplankton in contact with the spill. The fiddler crabs were probably affected more due to the fact that the oil remained on the surface and heavily oiled *Spartina* mud flats, soaked into the peat areas lining the shore, and entered the crab burrows.

If an organism can survive the initial high concentrations of both the oil and the WSF, then the next hurdle is to survive the sublethal concentrations. Maintenance of a viable invertebrate population depends both on the health of the ecosystem and on the ability of individual organisms to withstand the effects of sublethal concentrations of pollutants on their vital physiological processes. For crustaceans, the processes leading toward and occurring at ecdysis, as well as the process of limb regeneration, are critical and have been shown to be sensitive to the effects of a toxicant (Peddicord and McFarland 1976; Weis 1977, 1978; Fingerman and Fingerman 1978; Rao et al. 1979; Cantelmo et al. 1982; Callahan and Weis 1983; Wang and Stickle 1987; Weis et al. 1987; Clare and Costlow 1989). The length of the intermolt cycle, that is, the time between ecdysis and the next molt, may have important consequences, particularly for juvenile crabs that have a limited season for growth in northern latitudes. Crustaceans regenerate missing limbs prior to molting. In the Brachyura, the limb grows in a folded position within a layer of cuticle and unfolds at the next molt. Overall growth after each molt as well as the growth rate of regenerating limbs represents a useful measure of an organism's performance, since growth is an integral response to the environment, both functionally and temporally. It is also ecologically relevant, since it can be related to success of the population in the field (Sastry and Miller 1981).

The length of the intermolt cycle, growth after each ecdysis, and the regeneration of limbs are all affected by sublethal concentrations of environmental pollutants. Weis and Mantel (1976) found that exposure of regenerating fiddler crabs (*Uca pugilator*) to DDT produced an acceleration of regeneration and molting. This was attributed to the stimulatory effects of the organochlorine pesticide on the neuroendocrine system. Exposure of grass shrimp (*Paleomonetes pugio*) to sodium pentachlorophenate did not alter the duration of the intermolt cycle but caused a dose-related inhibition of limb regeneration (Rao et al. 1981).

Cristini, Cooper, and Bernard (1989) fed juvenile blue crabs clams harvested from the Elizabeth, New Jersey, mud flat prior to the Exxon spill. The crabs from Newark Bay had fewer molts; and those that molted exhibited a significant increase in the intermolt cycle (50 days to complete a molt vs. 40 days for controls). The animals (1 to 4 cm carapace width) fed Newark Bay clams grew only .27 centimenters, while control crabs grew .54 centimeters. The crabs fed Newark Bay clams exhibited a slower rate of limb regeneration.

No. 2 fuel oil is a refined oil and contains high concentrations of benzene, naphthalene, and methyl substituted naphthalene. These chemicals constitute a large portion of the water-soluble fraction. Cantelmo et al. (1982) reported that exposure of juvenile blue crabs, (*C. sapidus*), to 1.0 milligrams per liter of benzene increased the length of the intermolt cycle, decreased the growth increment per molt, and retarded the rate of limb regeneration. Animals exposed to benzene required an average of 50 days to complete the molt cycle, while control crabs required only 33 days. Control crabs (1 to 4 cm carapace width) grew an average of .46 centimeters at each molt, while crabs in the same size class treated with benzene grew an average of .18 centimeters per molt. Crabs exposed to benzene also exhibited a slower rate of growth of regenerating limb buds and a large plateau in the growth curve before molt. In addition, these crabs required a longer time to develop normal pigmentation in the regenerating limbs than did controls.

Both benzene and DMN have been shown to affect the rates of energy production in juveniles of this species at the level of the tissues and the whole organism (Cantelmo et al. 1981).

Cantelmo et al. (1982) also examined the effects of exposure of juvenile *C. sapidus* to .01 milligrams per liter of dimethylnaphthalene (DMN). The effects of DMN on length of the intermolt cycle are similar to those of benzene (51 days to complete a molt cycle). However, the growth increment at molt is only sightly reduced from that of the control (.37 cm). In addition, the pattern of limb bud regeneration is more like that of the controls despite the prolonged intermolt cycle.

Wang and Stickle (1987) reported that exposure of juvenile *C. sapidus* to the water-soluble fraction of "south Louisiana" crude oil inhibited growth and molting. Molting occurred twice in the crabs in their study. Growth, measured as percentage of size (carapace width) increased after molting, was significantly reduced in crabs exposed to 1.476 and 5.504 milligrams per liter of aromatic hydrocarbons with the first molt and at all hydrocarbon concentrations with the second molt. Growth during the second molt was inversely related to crude-oil levels, decreasing from 29.2 percent in control crabs to 16.9 percent in crabs exposed to 2.504 milligrams per liter of aromatic hydrocarbons. The intermolt period was prolonged in crabs exposed to crude oil and was positively correlated with exposure concentrations.

The data on frequency of molting and growth increment at molt suggest that juvenile blue crabs exposed to chemicals in their environment or in their food could grow less in a season than crabs in a clean environment. Clearly, data obtained in a laboratory study can be applied to populations in the field only with caution. However, the data on growth in these studies are within the range reported for this species in field studies (Darnell 1959; More 1969; Perry 1975). Cristini, Cooper, and Bernard (1989) have shown a reduction in size and alterations in processes associated with molt in animals fed food taken directly from polluted environments. This suggests that long-term exposure to chemicals in the environment could have detrimental effects on both individuals and populations of *C. sapidus*.

Summary of the Arthur Kill Spill Effects

The following conclusions can be drawn from both the literature and from observations made following the 1990 Exxon spill in the Arthur Kill. The invertebrates living on and in the mud flat were killed on contact with the oil. The animals that lived higher on the mud flat or were protected by rocks did survive. However many animals living higher on the mud flat continued to die for several months following the spill. By the late spring and early summer 1990 no histopathological lesions in soft-shelled clams could be attributed to the spill. There was evidence of recruitment of new spat and crustaceans on the mud flat in the summer of 1990. The most obvious effect that was still evident at that time was the reduced growth of *Spartina* and the release of oillike material from peat and from underneath rocks. The impact of the spill was not as well characterized as it could have been, largely because of the lack of a planned response for the spill. The fact that the spill oc- curred during the winter, when the majority of the animals inhabiting the salt marsh are in a dormant state or are not present, was fortunate as it minimized the effects of the spill on the wildlife and plant life (multicel- lular and unicellular) in the Kill.

Conclusions

We feel the following information should be obtained regarding a spill of toxic material in any estuary.

What is the type of oil spilled? This includes a list of characteristics of the oil in the water column (lighter oils float on the surface, while heavier crude oils generally sink and can cause mortality of benthic communities).

What is the characterization of the oil? A sample of the oil that was spilled should be obtained. This is important for following the finger- print of the oil and estimating the extent of toxicity that might occur.

What are current weather conditions? This has a major effect on the way the oil slick will behave.

What is the amount of oil spilled? How much was spilled and over what period of time, and what measures were taken to contain the spill?

Where did the spill occur? What is the volume of oil spilled in relationship to the area impacted.

Are any environmentally sensitive areas near the spill location? In spill areas with sensitive areas, geographical mapping of land use should be made and potential impact should be estimated by the company or industry prior to an accident. Contingency plans should be in place to protect these areas. Aerial photographs of the area should be on file for reference.

What sampling of water, sediment, and biota will be carried out? Samples of water at various depths should be taken as the slick progresses. Sediment and biota should also be sampled to correlate tissue burdens with exposure levels. In all cases the position of the sample and time and date are essential for evaluating impacts.

Who are the local researchers carrying out research in the spill area? A list of researchers and their interests should be compiled so that information gathered prior to the spill, during, and after can be assembled.

What types of organisms inhabit potentially impacted areas? A listing of the organisms in the areas and the densities of these organisms, if available, should be collected. Prespill evaluations on impacted areas are essential for evaluating the spill's total effect. Periodic surveys of organisms present and their density should be maintained by the company that has a potential spill situation. Standard sampling protocols should be established to prevent different groups collecting data that cannot be compared due to different sampling protocols.

What type of postspill cleanup should be used for a particular area? The types of cleanup procedures that should or should not be implemented should be determined in advance. This could be different for each type of habitat that might be impacted, but the use of dispersant can often increase the toxicity associated with spills. In the case of the Exxon Alaskan spill the use of steam cleaning for aesthetic purposes probably did more damage than leaving the oil.

What type of information should be gathered on the impacted invertebrate populations? Animals and the material in which they live should be sampled for chemical analysis. Invertebrates representing different trophic levels and feeding patterns should be sampled. Chemical analysis of organs can be useful in determining if levels reached high enough concentrations to have caused death. If levels in the whole organism are measured, the concentration of oil in target tissue will be diluted by organs with low levels of the material. The minimum amount of tissue for analysis depends on the material spilled, but a sample of 100 grams is normally sufficient. In any sample the minimum number of replicate samples is three, and more are better. If samples are pooled, then animals of similar size should be pooled.

The number of animals observed to be affected or dead should be noted. Because invertebrates decay very quickly in warm weather it is important to make observations within 24 hours following the spill. Behavioral alterations should be noted, as they could be related to possible histopathological effects. Animals should be collected for necropsy at both the grossly visible level and at the light-microscopic level. As a rule of thumb, if multiple samples are to be collected the number should be thirty individuals. If only a single sample will be obtained between sites, then two hundred should be examined. If animals are to be examined for histopathology, the tissue should be refrigerated, not frozen. Freezing destroys the cellular architecture of the tissue.

How often should samples be taken? Samples should be collected in populations outside the spill area at the time of the spill. Samples should be taken from populations that are impacted during the spill (time 0) and following the spill (24, 48, 72, and 96 hrs. postspill), and monthly until death occurs or lesions return to prespill levels.

Literature Cited

Anderson, J. W. 1973 (May 21–25). Uptake and depuration of specific hydrocarbons from oil by the bivalves *Rangea cuneata* and *Crassostrea virginica*. Workshop on Inputs, Fates and Effects of Petroleum in the Marine Environment, National Academy of Sciences, pp. 690–708.

————. 1975. *Laboratory Studies on the Effects on Oil on Marine Organisms: An Overview*. API publication no. 4249.

Berthou, F., G. Balouet, G. Bodennec, and M. Marchand. 1987. The occurrence of hydrocarbons and histopathological abnormalities in oysters for seven years following the wreck of the *Amoco Cadiz* in Brittany (France). *Mar. Envir. Res.* 23:103–133.

Brown, R. P., A. Cristini, and K. R. Cooper. 1992. Histopathological alterations in *Mya arenaria* following a #2 fuel oil spill in the Arthur Kill, Elizabeth, New Jersey. *Mar. Envir. Res.* 34:65–68.

Brown, R. S., and K. R. Cooper. 1978 (January 11–13). Histopathological analyses of benthic organisms from the vicinity of the *Argo Merchant* wreck.

Proceedings: In the Wake of the Argo Merchant, pp. 96–102. University of Rhode Island, Kingston.

Burger, J., J. Brzorad, and M. Gochfeld. 1991. Immediate effects of an oil spill on behavior of fiddler crabs (*Uca pugnax*). *Archiv. Environ. Contam. Toxicol.* 20:404–409.

Calahan, P., and J. S. Weis. 1983. Methylmercury effects on regeneration and ecdysis in fiddler crabs (*Uca pugilator, U. pugnax*) after short-term and chronic pre-exposure. *Arch. Environ. Contam. Toxicol.* 12:707–714.

Cantelmo, A., R. Lazell, and L. Mantel. 1981. The effects of benzene on molting and limb regeneration in juvenile *Callinectes sapidus. Mar. Biol. Lett.* 2:333–343.

Cantelmo, A., L. Mantell, R. Lazell, F. Hospod, E. Flynn, S. Goldberg, and M. Katz. 1982. The effects of benzene and dimethylnaphthalene on physiological processes in juveniles of the blue crab, *Callinectes sapidus*. In W. Vernberg, A. Calabrese, F.P. Thurberg, and F. J. Vernberg, eds., *Physiological Mechanisms of Marine Pollutant Toxicity*, pp. 349–389. New York: Academic Press.

Clare, A. S., and J. D. Costlow, Jr. 1989. Effects of the insecticide methomyl on development and regeneration in megalop and early postlarval stages of the mud crab, *Rhithropanopeus harrisii* (Gould). In D. Weigmann, ed., *Pesticides in Terrestrial and Aquatic Environments*, pp. 3–16. Blacksburg: Virginia Polytech. Inst. Water Resources Inst.

Cristini, A. 1987. An *in situ* study of adenylate energy charge and other biochemical parameters in the bivalve *Mya arenaria* from Raritan Bay and Long Island Sound. In W. B. Vernberg, A. Calabrese, F. Thurberg, and F. T. Vernberg, eds., *Pollution Physiology of Estuarine Organisms*, pp. 231–250. Columbia: University of South Carolina Press.

Cristini, A., and K. Cooper. 1992. The distribution of 2,3,7,8-tetrachlorodibenzo-p-dioxin in juvenile blue crabs, *Callinectes sapidus*, and in the physiological effects of consumption of food from a polluted environment on this species. In C. S. Walker and D. Livingston, eds., *Persistent Pollutants in Marine Ecosystems*, pp. 49–62. New York: Plenum.

Cristini, A., K. R. Cooper, and S. Bernard. 1989. The distribution of dioxins in the tissues of juvenile blue crabs, *Callinectes sapidus*. Society of Environmental Toxicology and Chemistry Tenth Annual Meeting, Toronto, Canada, Abstract 99.

Darnell, R. M. 1959. Studies of the life history of the blue crab (*Callinectes sapidus* Rathbun) in Louisiana waters. *Trans. Am. Fish. Soc.* 88:294–304.

Fingerman, S., and M. Fingerman. 1978. Effects of two polychlorinated biphenyls (aroclor 1242 and 1254) on limb regeneration in the fiddler crab *Uca pugilator*, at different times of the year. *Vie Milieu* 28–29:69–75.

Fossato, V. V. 1974. Elimination of hydrocarbons by mussels. *Mar. Biol.* 24:7–9.

Fossato, V. V. and W. J. Canzonier. 1976. Hydrocarbon uptake and loss by the mussel, *Mytilus edulis. Mar. Biol.* 36:243–250.

Galtstoff, P. S., H. F. Prytherch, R. O. Smith, and V. Koehring. 1935. Effects of crude oil pollution on oysters in Louisiana waters. *Bull. Bureau Fish., Wash. D.C.* 18:143–210.

Gardner, G. R., P. P. Yevich, and P. F. Rogerson. 1975. Morphological anomalies in adult oyster, scallop, and Atlantic silversides exposed to waste motor oil. *Conference on Prevention and Control of Oil Pollution:* 473–477.

Gilfillan, E. S. 1973 (March 13–15). Effects of seawater extracts of crude oil on carbon budgets in two species of mussels. *Proceedings Joint Conference on Prevention and Control of Oil Spills*, pp. 691–695. API, EPA, and USCG.

————. 1975. Decrease of net carbon flux in two species of mussels caused by extracts of crude oil. *Mar. Biol.* 29:53–57.

Gilfillan, E. S. and Vandermeulen. 1978. Alterations in growth and physiology of soft-shell clams *Mya arenaria* chronically oiled with Bunker C from Cheddbucto Bay, Nova Scotia, 1970–76. *Journal of Fisheries Research Board of Canada* 33: 680–686.

Hampson, G. R. and E. T. Moul. 1977. A three-year study at Winsor Cove, salt marsh grasses and #2 fuel oil. *Oceanus* 20:25–30.

James, M. O., and K. Shiverick. 1984. Cytochrome P-450-dependent oxidation of progesterone, testosterone and ecdysone in the spiny lobster, *Panulirus argus. Arch. Biochem. and Biophysics* 233:1–9.

Livingstone, D. R. 1985. Responses of the detoxication/toxication enzyme system of mollusks to organic pollutants and xenobiotics. *Mar. Pollut. Bulletin* 16:158–164.

More, W. R. 1969. A contribution to the biology of the blue crab (*Callinectes sapidus* Rathbun) in Texas, with a description of the fishery. *Tex. Parks Wildl. Dept. Tech. Serv.* 1:1–31.

Moreno, M. D., Cooper, K. R., and Georgopoulos, P. A. 1992. Physiologically based pharmacokinetic model for *Mya arenaria. Mar. Envir. Res.* 34:321–325.

Neff, J. M., B. A. Cox, D. Dixit, and J. W. Anderson. 1976. Accumulation and release of petroleum-derived aromatic hydrocarbons by four species of marine animals. *Mar. Biol.* 38:279–283.

Pauley, G. B., and A. K. Sparks. 1965. Preliminary observations on the acute inflammation reaction in the Pacific oyster, *Crassostrea gigas. J. Invert. Pathol.* 7:248–256.

Peddicord, R., and V. McFarland. 1976. Effects of suspended dredged material on the commercial crab, *Cancer magister.* In P. Krenkel, J. Harrison, and J. Burdick III, eds., *Dredging and Its Environmental Effects*, pp. 633–644. New York: American Society of Civil Engineers.

Perry, H. M. 1975. The blue crab fishery in Mississippi. *Gulf Res. Rep.* 5:39–57.

Polak, R., A. Filion, S. Fortier, J. Lanier, and K. Cooper. 1978 (January 11–13). Observations on *Argo Merchant* oil in zooplankton of Nantucket shoals. *Proceedings: In the Wake of the* Argo Merchant, pp. 109–115. University of Rhode Island, Kingston.

Prince, R. and K. R. Cooper. 1992. Decreased sensitivity to 2,3,7,8-tetrachloro-dibenzo-p-dioxin in a TCDD-impacted *Funudulu heteroclitus* population compared to a TCDD-nonimpacted population. *The Toxicologist* 12:1:82.

Rao, K. R., F. R. Fox, P. J. Conklin, and A. C. Cantelmo. 1981. Comparative toxicology and pharmacology of chlorophenols: Studies on the grass shrimp, *Palaemonetes pugio.* In F. J. Vernberg, A. Calabrese, F. P. Thurberg, and W. B. Vernberg, eds., *Biological Monitoring of Marine Pollutants,* pp. 37–72. New York: Academic Press.

Rao, K. R., F. R. Fox, P. J. Conklin, A. C. Cantelmo, and A. C. Brannon. 1979. Pological and biochemical investigations of the toxicity of pentachlorophenol to crustaceans. In W. B. Vernberg, F. P. Thurberg, A. Calabrese, and F. J. Vernberg, eds., *Marine Pollution: Functional Responses,* pp. 307–340. New York: Academic Press.

Sastry, A. N., and D. C. Miller. 1981. Application of biochemical and physiological responses to water quality monitoring. In F.J. Vernberg, Calabrese, Thurberg, and W.B. Vernberg, *Biological Monitoring,* pp. 265–294.

Singer, S. C., and R. F. Lee. 1977. Mixed function oxygenase activity in blue crab, *Callinectes sapidus:* Tissue distribution and correlation with changes during molting and development. *Biol. Bull* 153:377–386.

Sparks, A. K. 1972. Reaction to injury and wound repair in invertebrates. In *Invertebrate Pathology,* pp. 58–108. New York and London: Academic Press.

Squibb, K. S., J. M. O'Connor, and T. J. Kneip. 1991. New York/New Jersey harbor estuary toxics categorization. Prepared for U.S. Environmental Protection Agency Region II, New York, New York.

Stainken, D. M. 1976. A descriptive evaluation of the effects of No. 2 fuel oil on the tissues of the soft shell clam, *Mya arenaria* L. *Bull. Environ. Contam. Toxicol.* 16:730–738.

Stainken D. M. 1978. Effects of Uptake and Discharge of Petroleum Hydrocarbons on the Respiration of the soft-shell clam, *Mya arenaria. J. Fish Res. Board of Canada* 35(5):637–642.

Stegeman, J. G. and J. M. Teal. 1973. Accumulation, release, and retention of petroleum hydrocarbons by the oyster, *Crassostrea virginica. Mar. Biol.* 22:37–44.

Thurberg, F. P., E. Gould, and M. A. Dawson. 1978 (January 11–13). Some physiological effects of the *Argo Merchant* oil spill on several marine teleosts and bivalve mollusks. *Proceedings: In the Wake of the* Argo Merchant, pp. 103–108. University of Rhode Island, Kingston.

Wang, S. Y., and W. B. Stickle. 1987. Bioenergetics, growth and molting of the blue crab, *Callinectes sapidus,* exposed to the water-soluble fraction of South Louisiana crude oil. In W. B. Vernberg, Calabrese, Thurberg, and F. J. Vernberg, *Pollution Physiology,* pp. 107–126.

Weis, J. S. 1977. Limb regeneration in fiddler crabs: Species differences and effects of methylmercury. *Biol. Bull.* 152:263–274.

———. 1978. Interactions of methylmercury, cadmium, and salinity on regeneration in the fiddler crabs *Uca pugilator, Uca pugnax,* and *Uca minax. Mar. Biol.* 49:119–124.

Weis, J. S., J. Gottlieb, and J. Kwiatkowski. 1987. Tributyltin retards regeneration and produces deformities of limbs in the fiddler crab *Uca pugilator. Arch. Environ. Contam. Toxicol.* 16:321–326.

Weis, J. S., and L. H. Mantel. 1976. DDT as an accelerator of regeneration and molting in fiddler crabs. *Estuarine Coastal Mar. Sci.* 4:461–466.

Weis, J. S. and P. Weis. 1989. Effects of environmental pollutants on early fish development. *CRC Critical Reviews, Aquatic Sciences* 1:45–73.

Weis, J. S., P. Weis, M. Heber, and S. Vaidya. 1981. Methylmercury tolerance of killifish (*Fundulus heteroclitus*) embryos from a polluted vs. non-polluted environment. *Marine Biology* 65:283–287.

————. 1982. Investigations into mechanisms of heavy metal tolerance in killifish (*Fundulus hetroclitus*) embryos. In W. B. Vernberg, Calabrese, Thurberg, and F. J. Vernberg, *Physiological Mechanisms*, pp. 311–330.

10. Fiddler Crabs (*Uca* spp.) as Bioindicators for Oil Spills

Joanna Burger,
John N. Brzorad, and
Michael Gochfeld

Government, the private sector, and the public are increasingly becoming aware that society needs to understand our native ecosystems and to evaluate their health and well-being. The public clearly perceives some environmental degradation as risks they are unwilling to accept (Slovic 1987; Fischer et al. 1991). It is essential not only to assess damage following an environmental insult but to track changes in natural communities, habitats, and ecosystems over time. Measuring changes in environmental quality usually involves monitoring air, soil, and water quality, and monitoring individuals and populations of living organisms (biomonitoring) (Karr 1991). Given that time and resources are likely to be limited, it is not possible to monitor all aspects of the air, soil, water, and biota continuously, or even periodically. Thus, some subset of the physical and biotic components or functions are usually monitored on a regular basis.

This chapter discusses physical and biological monitoring, describes the use of indicator species for biomonitoring, and uses research on fiddler crabs (*Uca* spp.) in the Arthur Kill following the 1990 Exxon oil spill as an example of bioindicators.

Environmental Monitoring

Assessing environmental quality is essential for managing and regulating our environment from a human-health viewpoint, as well as from an ecological-health perspective. Evaluating the threats to human health, often referred to as environmental- or human-risk assessment, is a well-established (if controversial) procedure. The paradigm and methodology for evaluating risks to humans were set forth in 1983 by the National Research Council (National Research Council 1983) of the National Academy of Sciences. Human-risk assessment normally involves four phases: hazard identification, dose–response analysis, exposure assessment, and risk characterization. This formalization ensures some uniformity in these evaluations, as well as encourages attention at all levels of human health risk.

The assessment of risks to species, communities, and ecosystems (ecological-risk assessment) has proceeded along different lines, and only recently has a formalized methodology begun to emerge. Assessment of ecological risks has usually involved either examining potential damage, the results of which are reported in environmental impact statements, or focusing on threatened and endangered species. Both aspects have their basis in environmental regulations that ensure that appropriate procedures have been conducted before new developments or changes in existing facilities occur.

Recently, considerable attention has been devoted to the methodology of ecological-risk assessment (Suter, Barnthouse, and O'Neill 1987; Cairns 1990; Hunsaker et al. 1990; Suter 1990). Ecological-risk assessment has emerged as a critical method for understanding local as well as global phenomena (Mooney 1991; Ojima et al. 1991). Ecological assessment covers a range from single species to ecosystem effects, and from local to global concerns (Sheehan et al. 1984). The examination of effects of pollutants on single species has been well developed and constitutes the field of ecotoxicology (Bascietto et al. 1990; Burton 1990; Okkerman et al. 1991). Some scientists have extended ecotoxicology to include the effects of pollution on ecosystems as well (Harris et al. 1990).

Basic to any ecological-risk assessment is an inventory of the existing organisms; an assessment of their health, behavior, interactions, and population levels; and an understanding of the structure and function of the existing populations, communities, and ecosystems (Burger and Gochfeld 1992). Accomplishing this for most habitats entails basic ecological research on the structure and function of ecosystems (with their component parts) as well as monitoring of these systems to assess status and changes within the system.

Identifying the status and trends in biological systems requires the selection of component parts or functions (i.e., nutrient cycling, primary productivity) for monitoring, since one cannot monitor every organism or every function of an ecosystem. An important step in monitoring is selecting indicator organisms or functions.

Features that are usually considered in selecting a bioindicator species include (1) the species's widespread persistence in the system being examined, (2) its sensitivity to change, (3) its usefulness to managers and regulators, (4) its importance to the public, (5) acceptable methods for its assessment, (6) characteristic of the ecosystem, and (7) economic feasibility. Although not exhaustive, this list illustrates the importance of considering biological, societal and economic concerns when choosing an indicator species.

The need for biomonitoring to assess anthropogenic impacts on the global environment is growing. New programs have been initiated, such as the International Geosphere Biosphere Program, which will conduct long-term monitoring of a network of sites on a worldwide basis (Brun,

Wiersma, and Rykiel 1991). Sites forming the focal point for physical, chemical, and biological measurements will represent major biomes (SCOPE/MAB 1987). These programs will be a key component of evaluating global environmental change.

In the United States several state and federal agencies have initiated biomonitoring schemes and have selected specific species as bioindicators. The National Oceanic and Atmospheric Administration (NOAA) has developed a National Status and Trends Program that includes several projects such as benthic surveillance (National Oceanic and Atmospheric Administration 1988) and mussel watch (National Oceanic and Atmospheric Administration 1989), as well as chemical contamination (O'Connor and Ehler 1991). The U.S. Fish and Wildlife Service is conducting a toxics-monitoring program that examines contaminants in wings of ducks and starlings (*Sturnus vulgaris*). The Environmental Protection Agency has initiated the Environmental Monitoring and Assessment Program (EMAP) to produce annual statistical reports and interpretive summaries that will identify adverse disturbances. An example of a regional biomonitoring scheme for contaminants is one being conducted in the Great Lakes, where the herring gull (*Larus argentatus*) is used to monitor levels and effects of organochlorine contamination (Mineau et. al. 1984) and ecotoxicology tests are used to examine other contaminants (Harris et al. 1990).

Most states have toxics-monitoring systems that assess water quality on a regular basis. Karr (1991) examined methods of assessing water quality by defining biological degradation of supplies of water and measuring biotic integrity. A number of state agencies have begun biomonitoring programs for species that are threatened, endangered, or of special concern (see Burger 1990; Downer and Liebelt 1990; Burger et al. 1994). Some private agencies, such as the Nature Conservancy, also monitor populations and reproductive success for species of concern.

Basic to most of these monitoring schemes is the selection of indicator species that can be used over many years to assess changes. Once the indicator species are selected, then measurable variables must be selected. These might include disease or injury prevalence, size or growth rates, reproductive rates, recruitment rates, behavior, age structure of the population, or dispersion in time and space. Our ability to monitor these factors varies for different species. For example, in birds current population size and reproductive success can be used effectively to predict future population levels (see Burger and Gochfeld 1991), whereas for fish knowing recruitment and growth rates in addition to current population size may be essential.

A number of organisms have been used as bioindicators for marine and coastal systems, including mussels, mainly the blue mussel (*Mytilus edulis*) (National Oceanic and Atmospheric Administration 1989), and marine birds (Peakall 1991; Nisbet 1994). Mussels and birds are different in the following ways:

Mussels live in one place, whereas birds may migrate thousands of kilometers from where they breed.

Mussels live for only a few years, whereas marine birds may live for thirty or forty years.

Mussels have a high fecundity, whereas most marine birds produce three or fewer young per year.

Mussels are sessile and feed in one small area, whereas marine birds may feed over large areas around their nesting colonies.

Mussels provide few opportunities for the study of complex behavior, whereas birds provide many.

These differences mean that both types of species are useful bioindicators because they can tell us different things about the environment. Mussels indicate pollution at one point in time, at one place; birds integrate exposure over time and space. Individual mussels can be followed for only a few years, whereas the same birds can be followed over many years, allowing an examination of bioaccumulation and providing a model for higher-level consumers such as humans.

The next section explores the use of fiddler crabs (*Uca* spp.) as bioindicators for marine systems. Fiddler crabs fall between the mussel and bird models in mobility and in complexity of behavior.

Fiddler Crabs as Bioindicators

Fiddler crabs of the genus *Uca* are widespread and abundant in estuarine habitats around the world (Crane 1975). The marsh fiddler crab (*Uca pugnax*) occurs in eastern North America from Cape Cod to Florida (Ward, Howes, and Ludwig 1976) and provides an important food source for organisms at higher trophic levels, such as birds and mammals. These crabs are relatively sedentary and are susceptible to environmental pollutants such as metals (Decoursey and Verberg 1972), polychlorinated biphenyls (Nimmo et al. 1971; Krebs et al. 1974), and insecticides (War and Busch 1976; Vernberg, Guram, and Savory 1977).

The characteristics that make fiddler crabs excellent bioindicators are their widespread distribution geographically, abundance (assuring that they can be monitored), role in the food chain, sedentary habits (making them vulnerable to toxics and indicative of local exposure), and documented responses to environmental pollutants. Furthermore, fiddler crabs engage in a wide variety of behaviors above ground, allowing an examination of the sublethal effects of toxics on behaviors that affect survival and reproductive success. In the active season (April to October in the northeastern United States), fiddler crabs normally feed on the surface when the tide is low and mud flats are exposed. As the tide rises the crabs enter underground burrows and cover the opening with a mud plug. While active on the surface fiddler crabs explore the mud surface, feed on surface algae, defend territories, court, and engage in sexual activity. This wide array of behaviors allows evaluation of toxic effects. In the winter the crabs remain about 0.3 to 1.5 meters below ground in a tunnel with a mud plug near the surface.

Fiddler Crabs and Oil Spills

The effects of oil spills on fiddler crabs have been examined in a number of places around the world. In the Niger delta of Nigeria, Snowden and Ekweozor (1987) reported initial mortality to *Uca tangeri* but found no significant population effects. Similarly, initial mortality of *U. pugnax* was reported following an oil spill at West Falmouth, Massachusetts (Krebs and Burns 1977). At West Falmouth, however, the crab populations also suffered long-term effects including reduced species diversity, reduced ratio of females to males, reduced juvenile settlement, heavy overwinter mortality, incorporation of oil into body tissues, locomotory impairment, and abnormal burrow construction. The overwinter mortality was partially the result of the faulty burrow construction. In the winters following the oil spill, some crabs built burrows that were too shallow, and they died during the winter. The population effects on the fiddler crabs were still evident 7 years after the spill (Krebs and Burns 1977).

The studies conducted at West Falmouth on fiddler crabs were initiated $2\frac{1}{2}$ years after the oil spill. The immediate and short-term effects of an oil spill were not examined. Thus, the 1990 oil spill in the Arthur Kill provided a unique opportunity not only to assess immediate damage (see Chapter 7) but to examine short-term effects. As documented, immediate mortality followed the oil spill (Burger, Brzorad, and Gochfeld 1991, 1992). Although the Arthur Kill oil spill occurred during the winter, when the crabs were well below ground, the oil forced many crabs to the surface. Many ultimately died in pools of oily water (Burger, Brzorad, and Gochfeld 1992).

Fiddler Crabs and the Arthur Kill Spill

The effects of an oil spill on fiddler crabs are of interest because of the population dynamics and survival of the crabs. However, changes in and contamination of fiddler crab population are also important because of their role in the food chain. This aspect makes them particularly useful as bioindicators of environmental problems.

Some twelve hundred pairs of colonial waterbirds breed on the three islands in the Arthur Kill, along with thriving colonies of gulls (*Larus* spp.) and expanding colonies of double-crested cormorants (*Phalacrocorax auritus*) (Parsons, Maccarone, and Brzorad 1991). These birds feed on many organisms in the Kill, including fish, shrimp, fiddler crabs, and a variety of other invertebrates. Herons, egrets, and ibises feed primarily in the tidal creeks and marshes, and along the river edges, although some species also feed on nearby landfills. Yellow-crowned night herons (*Nyctanassa violacea*), endangered in both New York and New Jersey, feed almost exclusively on fiddler crabs. Any reduction in the quantity or quality of the food supply could depress reproductive success by lowering the survival of the young birds. Furthermore, population levels will decline not only because of lowered reproductive success but because adults that fail to raise young often switch colony sites (Burger 1982).

Our study was designed to examine the population levels and behavior of fiddler crabs at five study sites at different distances from the 1990 oil spill and to document the availability of crabs as a food source for other organisms.

Field Studies with Fiddler Crabs

Methods

Fiddler crabs at four study sites (Fig. 10.1) along the Arthur Kill, and one at nearby Cheesequake marsh were examined. Cheesequake was used as a control site because it is a similar habitat to that of the Arthur Kill in a nearby, relatively clean salt marsh. All five study sites were located on tidal mud-flat banks along tidal creeks. Thus, the creek bed bordered one side, and a fringe of cordgrass (*Spartina alterniflora*) bordered the other.

Each of the five sites was studied for 1 day each week from April 15 to August 15, 1990 (Fig. 10.2). Usually data could be gathered for 5 to 6 hours a day during low tide. To reduce disturbances, field personnel arrived before the mud flat was uncovered, when all crabs were still underground in their burrows.

The sampling regime consisted of frequency sampling on each study plot. Before sampling, the date, time of day, and hours since low tide were

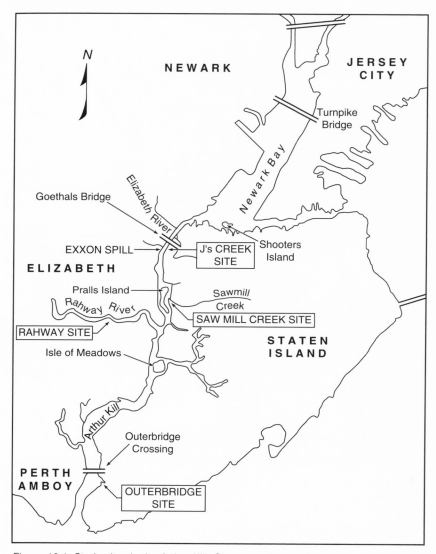

Figure 10.1. Study sites in the Arthur Kill. Cheesequake not shown.

recorded. At 15-minute intervals the presence, location, and activity of all crabs within the observation plot by species (*Uca pugnax, U. minax*) and by sex (male fiddler crabs are easily identified by their large claw, used for courtship and territorial defense) were recorded. The behaviors record-ded included the number of crabs in the plot that were entirely out of their

Figure 10.2. John Brzorad and Joanna Burger collecting dead and live emerged crabs in January 1990.

burrow, partially out of their burrow, feeding, fighting, moving, displaying, and stationary. The usual protocol was for one field assistant to record behavior, while the second used binoculars to scan the banks, verbally noting the number of crabs engaging in each activity (by sex and species).

Results

A GLM (general linear model) regression procedure (SAS 1985) was used to examine factors that contribute to or explain variations in the number of fiddler crabs and their holes, and their behavior (Table 10.1).

Table 10.1. Models Explaining Variations in the Number of Fiddler Crabs, Crab Holes, and Crab Behavior

	No. of holes	No. of U. pugnax	No. of U. minax	% moving	% feeding	% stationary
Model						
F	317.8	49.0	80.7	5.7	12.5	41.0
r^2	.79	.28	.39	.60	.12	.30
$P <$	.0001	.0001	.0001	.0001	.0001	.0001
Factors entering the model (F values/P values)						
Date	1575 (.0001)	53.6 (.0001)	163.1 (.0001)	7.02 (.008)	94.3 (.0001)	415.8 (.0001)
Study site	853 (.0001)	102.4 (.0001)	196.1 (.0001)	4.9 (.0006)	2.96 (.01)	18.5 (.0001)
Air temperature	—	6.6 (.01)	42.7 (.0001)	7.5 (.006)		10.1 (.001)
Water temperature	3.6 (.05)	24.1 (.0001)	6.32 (.01)	—		
Tide time	4.3 (.03)	27.1 (.0001)	19.1 (.0001)	—	23.3 (.0001)	27.9 (.0001)
Day time	—	12.2	47.9	—	13.9	11.4
Number of other Uca species	—	214.7 (.0001)	214.5 (.0001)	—		

The GLM procedure is a method of determining what factors affect fiddler crabs, since environmental conditions vary from day to day and collecting the data from all sites at the same time or under the same temperature regimes is difficult.

The number of fiddler crabs of both *Uca pugnax* and *U. minax* that were present was affected by date, study site, air and water temperatures, tide, time of day, and the number of other crabs present. Date, study site, water temperature, and tide also affected the number of crab holes present. Tide refers to the number of hours since low tide and is a measure of the time the mud flat has been exposed. When the mud flat is covered by water the fiddler crabs remain in their underground burrows with the entrances plugged with mud.

The only variables that affected all behaviors were date and study site, although tide, temperature, and time of day affected some (Table 10.1). Fiddler crabs are more active as the temperature warms up. The models suggest not only that environmental variables affected the number of fiddler crabs present and their behavior, but study site itself affected all aspects of the abundance and behavior of fiddler crabs following the 1990 oil spill.

When the overall average characteristics from the five study sites were examined, temperature was found to vary significantly, by as much as 16°C, a substantial difference, among the sites. This was due in part to the difficulties of reaching some of the study sites early in the season and in part to normal temperature variation. Because of these sampling problems, and because the models from the entire data set indicated that temperature was consistently one of the most important factors affecting the abundance and behavior of fiddler crabs, the decision was made to examine a 1-week period when temperatures were fairly similar at all sites as an indication of the whole sample period (Table 10.2).

Even within any given week differences in air and water temperatures existed. Some of the variability in temperature is site specific and is superimposed on changes associated with differing times of low tide on succeeding days. Water temperature was always 2 to 3°C below air temperatures at all sites (Table 10.2).

Overall, the number of *Uca pugnax* was positively correlated with the number of *U. minax* at all sites (Table 10.3). Where there were many of one species, there were also many of other species. The number of *U. pugnax* was inversely related to tide, or mud-flat exposure. This suggests that more crabs came out early, when the tide first receded, and the number declined a few hours later. This is not surprising, since they had been inactive in their burrows during high tide and

Table 10.2. Comparison of Study Sites during 1 Week with Comparable Temperature

	J's Ditch	Sawmill Creek	Rahway	Outerbridge	Cheesequake	Kruskal-Wallis X^2 (p)
Air temperature (°C)	29 ± 0	28 ± 0	25 ± 0	26 ± 0	30 ± 0	407 (.0001)
Water temperature (°C)	27 ± 0	25 ± 0	23 ± 0	24 ± 0	27 ± 0	407 (.0001)
Tide time	1.5 ± 0	1.5 ± 0	1.5 ± 0	1.6 ± 0	1.5 ± 0	NS
Day time	12.12 ± 0.2	15.19 ± 0.2	12.20 ± 0.2	12.23 ± 0.2	12.12 ± 0.2	154 (.0001)
No. of holes	14.4 ± 0.4	16.6 ± 0.3	30 ± 1.0	29 ± 1.1	22 ± 3	285 (.0001)
Total *Uca*	11.8 ± 0.2	10.4 ± 0.3	6.7 ± 0.4	11.3 ± 0.5	11.5 ± 0.4	120 (.0001)
No. of *U. pugnax*	10.9 ± 0.7	9.2 ± 0.3	2.5 ± 0.3	10.6 ± 0.7	6 ± 0.5	120 (.0001)
No. of *U. minax*	0.6 ± 0.1	0.5 ± 0.6	3.9 ± 0.5	0.4 ± 0.1	5.3 ± 0.4	120 (.0001)
% feeding	24 ± 2	41 ± 0	26 ± 3	34 ± 2	42 ± 1	50 (.0001)
% moving	12 ± 2	5 ± 1	2 ± 0	9 ± 1	7 ± 1	47 (.0001)
% fighting	2 ± 0	0 ± 0	0 ± 0	0 ± 0	0 ± 0	17.2 (.001)
% displaying	15 ± 2	30 ± 2	9 ± 2	19 ± 3	34 ± 1	82.8 (.0001)
% stationary	46 ± 3	23 ± 2	63 ± 4	36 ± 0	16 ± 2	106 (.0001)
Ratio *Uca*/holes	.80	.60	.22	.37	.52	139 (.0001)

NOTES: Mean ± standard error.
0 designates standard error < .1. NS = not significant.

Table 10.3. Correlation among Numbers of *Uca* Species and Tides in the Different Study Sites

	No. of *U. pugnax* and *U. minax*	Tide time and No. of *U. pugnax*	Tide time and No. of *U. minax*
J's Creek	.10 (.01)	−.21 (0.0001)	.08 (NS)
Sawmill Creek	.09 (.02)	−.11 (0.008)	−.03 (NS)
Rahway	.58 (.001)	−.39 (0.001)	−.29 (0.0001)
Outerbridge	.13 (.006)	−.23 (0.001)	−.18 (0.0001)
Cheesequake	.71 (.0001)	−.06 (NS)	−.05 (NS)

NOTE: Given are Kendall tau correlations (probability levels).

were no doubt hungry. The relationship was similar, but not as strong, for *U. minax* (Table 10.3).

In examining the abundance data, it is clear that no differences existed in the total number of fiddler crabs present in the plots, except for the Rahway site. This is not surprising, since the Rahway site, on the Rahway River, is exposed to lower-salinity water, whereas all the other sites were salt marsh creeks without a source of freshwater inundation. Thus, the remaining comparisons in this chapter exclude the Rahway site. Furthermore, there has been some suggestion that industrial dumping of heavy metals is taking place at the mouth of the Rahway, which may also account for these differences.

The number of fiddler crabs was similar among the four sites (excluding Rahway), indicating no obvious effect of the oil spill on crab abundance. Nonetheless, it is instructive to examine the behavior of the crabs (Table 10.2). The number of holes or crab burrows in the study plots varied significantly. Fewer holes were found in the two sites closest to the oil spill (J's Creek, Sawmill Creek) than in the two sites farthest from the spill (Outerbridge, Cheesequake). Similarly, all aspects of their behavior varied significantly among the sites (Fig. 10.3). More crabs were stationary at J's Creek, immediately across from the oil spill, than at the other sites. Thus more crabs were vulnerable to predators and were not engaged in sexual, territorial, or feeding activities necessary for their reproduction and survival. Given the relative inactivity of the fiddler crabs at J's Creek, it was surprising that they engaged in significantly more aggression than the crabs at any other site. Since density was the same, a difference in aggression could not be attributed to increased encounter rates.

The relative inactivity of crabs at the sites closest to the spill suggests the possibility that population sizes may have been lower here. But a difference could be masked if a higher percentage of crabs remained on

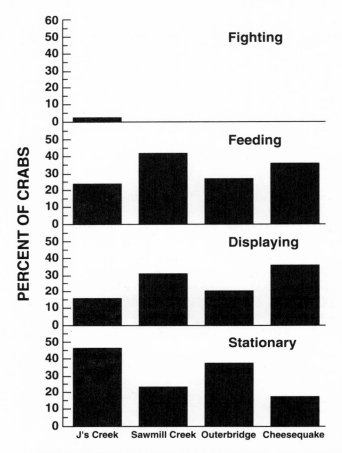

INCREASING DISTANCE FROM OIL SPILL

Figure 10.3. Percentage of crabs engaged in several behaviors at four salt marsh sites.

the surface, inactive, rather than going in and out of the burrows. To examine this possibility, individual crabs should be marked and population size determined by a capture/recapture method.

Overall the analysis showed that the density and abundance of crabs at the four salt marsh creek sites (exclusive of the Rahway River) did not vary as a function of distance from the oil spill. However, the behavior of the fiddler crabs did vary significantly. Fiddler crabs in the creek immediately across from the oil spill were less active, yet those that were active engaged in more aggression than crabs at all the other sites.

Discussion

In the best of all worlds, it is ideal to have similar data contrasting behavior and abundance before and after an oil spill. Such data are available for fish, shrimp, and birds in the Arthur Kill region (see Chapters 11–13). Another method for examining any perturbation to an ecosystem, however, is to compare abundances and behavior at different distances from the event. If the perturbation has an effect, then the response measures should show significantly different levels near the spill.

This comparative approach was used to examine the possible effects of the 1990 oil spill on fiddler crabs in the Arthur Kill. The results suggest that abundance did not vary as a function of the oil spill; however, behavioral changes were significant. The fiddler crabs at J's Creek, immediately across from the spill, were less active than the crabs at the other sites, and they engaged in significantly more aggression.

Although abundance did not vary among the four salt marsh creek sites, without previous censuses it is impossible to be sure that population size was not affected. But at present no evidence suggests an effect on population sizes 6 to 8 months after the spill. Evidence does, however, suggest altered behavior as a possible effect of exposure to oil. The effect may of course be due to the massive 1990 Exxon oil spill, or it could result from the cumulative effects of several spills that occurred in 1990.

The fiddler crabs at J's Creek were more often stationary than the crabs at the other creeks farther away from the spill. This has implications for the well-being and survival of the fiddler crabs. These crabs spend less time feeding than the other crabs did, suggesting that they may not be as healthy, or grow and molt as fast, as crabs at the other sites. Presumably, size relates directly to the crabs' ability to defend territories, compete for females, and avoid predators. The crabs also made fewer burrows per crab at J's Creek than at the other creeks. Having many holes is adaptive because it provides all crabs with close havens when predators approach.

Conclusions

Our experiences with studying the fiddler crabs in the Arthur Kill have suggested that, whenever possible, it is wise to have baseline data on abundance, distribution, and behavior of organisms. There is no substitute for information on undisturbed population levels and behavior. The most compelling case can be made for responses to a perturbation (an oil spill or other event) when there is both pre- and postevent data. Only then can direct comparisons be made.

Failing this, the comparative approach used here, in conjunction with multivariate statistical procedures that can separate the effects of environmental variables, can be effective in ascertaining the effects of an oil spill or other toxic event. Using this approach, it is essential to select study sites that are as similar as possible with respect to physical features (tidal regime, salinity, creek size, and location, mud-flat space, graduation or slope of mud flat) and biotic features (species and density of vegetation, faunal community structure). Any information about the target species, in this case the fiddler crabs, should also be used to select sites that are as similar as possible with respect to physical and biological properties.

Physical constraints such as access to the sites and the logistical problems associated with tides should be considered in selecting study sites. Whenever possible, it is ideal to have study sites that do not require boat transport. Optimally, study sites should be located on public land, where access can be assured for many years.

In most cases the sample design should include sampling immediately after the spill, for some months thereafter, and for some years thereafter. As the habitat and communities recover, the effects of the spill can then be further examined. For example, although no evidence of differences in abundance among study sites (when temperatures were the same) was found, if after several years differences emerge, then these initial conclusions might require reevaluation. Differences in behaviors, such as foraging, could lead to lower egg production and population declines in the future.

One important aspect to bear in mind is that if conducted long enough, studies following an oil spill can serve as baseline information for future events. Ecological or human-induced perturbations are not isolated events, and the probability of a future event is relatively high. This is especially true for areas such as the Arthur Kill, with its high density of oil storage and transfer facilities.

Acknowledgments

This project was funded by a cooperative contract from the New Jersey Department of Environmental Protection and the New York Department of Environmental Conservation through Louis Berger & Associates, Inc., and was aided by the Environmental and Occupational Health Sciences Institute (grant ES 05022 from NIEHS). We thank Kelly McFadden, Gil Cardoso, Leo Tsao, and Ralph Hua for field assistance.

Literature Cited

Bascietto, J., D. Hinckley, J. Piafkin, and M. Slimak. 1990. Ecotoxicity and ecological risk assessment: Regulatory applications at EPA. *Environ. Sci. Technol.* 24:10–15.

Brun, D. A., G. A. Wiersma, and E. J. Rykiel, Jr. 1991. Ecosystem monitoring at global baseline sites. *Environ. Monit. Assess.* 17:3–31.

Burger, J. 1982. The role of reproductive success in colony-site selection and abandonment in black skimmers *Rynchops niger. Auk* 99:109–115.

———. 1990. Least tern and black skimmer surveys of New Jersey 1989–1990. Reports to New Jersey Department of Environmental Protection, Trenton, N.J.

Burger, J., J. Brzorad, and M. Gochfeld. 1991. Immediate effects of an oil spill on behavior of fiddler crabs *Uca pugnax. Arch. Environ. Contam. Toxicol.* 20:404–409.

———. 1992. Effects of an oil spill on emergence and mortality in fiddler crabs *Uca pugnax. Environ. Monit. Assess.* 22:107–115.

Burger, J., and M. Gochfeld. 1991. *The Common Tern: Breeding Behavior and Biology.* New York: Columbia University Press.

———. 1992. Temporal scales in ecological risk assessment. *Arch. Environ. Cont. Toxicol.* 233:484–488.

Burger, J., K. Parsons, D. Wartenberg, C. Safina, J. O'Connor, and M. Gochfeld. 1994. Biomonitoring using least terns, common terns and black skimmers in the northeastern United States. *Coastal Res.* 10:39–47.

Burton, G. A. 1990. Ecotoxicology. *Environ. Sci. Technol.* 24:9.

Cairns, J. 1990. The genesis of biomonitoring in aquatic ecosystems. *Environ. Professional* 12:169–176.

Crane, J. 1975. *Fiddler Crabs of the World.* Princeton, N.J.: Princeton University Press.

Decoursey, P. J., and W. B. Vernberg. 1972. Effect of mercury on survival, metabolism and behavior of larval *Uca pugilator* (Brachyura). *Oikos* 23:241–247.

Downer, R. H., and C. E. Liebelt. 1990. *1989 Long Island Colonial Waterbird and Piping Plover Survey.* Islip: New York Department of Environmental Conservation.

Fischer, G. W., M. G. Morgan, B. Fischhoff, I. Nair, and L. B. Lave. 1991. What risks are people concerned about? *Risk Analysis* 11:303–315.

Harris, H. J., P. E. Sager, H. A. Regier, and G. R. Francis. 1990. Ecotoxicology and ecosystem integrity: The Great Lakes examined. *Environ. Sci. Technol.* 24:598–603.

Hunsaker, C. T., R. L. Graham, G. W. Suter, R. V. O'Neill, L. W. Barnthouse, and R. H. Gardner. 1990. Assessing ecological risk on a regional scale. *Environ. Managem.* 14:325–332.

Karr, J. R. 1991. Biological integrity: A long-neglected aspect of water resource management. *Ecol. Applications* 1:66–84.

Krebs, C. T., and K. A. Burns. 1977. Long-term effects of an oil spill on populations of the salt-marsh crab *Uca pugnax*. *Science* 197:484–487.

Krebs, C. T., I. Valilla, G. R. Harvey, and J. M. Teal. 1974. Reduction of field populations of fiddler crabs by uptake of chlorinated hydrocarbons. *Marine Poll. Bull.* 5:140–142.

Mineau, P., G. A. Fox, R. J. Norstrom, D. V. Weseloh, D. J. Hallett, and J. A. Ellenton. 1984. Using the herring gull to monitor levels and effects of organochlorine contamination in the Canadian Great Lakes. In J. O. Nriagu and M. S. Simmons, eds., *Toxic Contaminants in the Great Lakes*, pp. 425–452. New York: Wiley.

Mooney, H. A. 1991. Biological response to climate change: An agenda for research. *Ecol. Applic.* 1:112–117.

National Oceanic and Atmospheric Administration. 1988. *Benthic Surveillance Project*. Rockville, Md.: National Oceanic and Atmospheric Administration, Ocean Assessments Division.

———. 1989. A summary of data on tissue contamination from the first three years (1986–1988) of the mussel watch project. Rockville, Md.: National Oceanic and Atmospheric Administration.

National Research Council. 1983. *Risk Assessment in the Federal Government: Managing the Process*. Washington, D.C.: National Academy Press.

Nimmo, D. R., P. D. Wilson, R. R. Blackman, and A. J. Wilson. 1971. Polychlorinated biphenyl absorbed from sediments by fiddler crabs and pink shrimp. *Nature* 231:50–52.

Nisbet, I.C.T. 1994. Effects of pollutants on marine birds. In D. H. Nettleship, J. Burger, and M. Gochfeld, eds., *Seabirds on Islands: Threats, Case Studies and Action Plans*. Cambridge, England: International Council for Bird Preservation.

O'Connor, J. P., and C. N. Ehler. 1991. Results from the NOAA national status and trends program on distribution and effects of chemical contamination in the coastal and estuarine United States. *Environ. Monit. Assess.* 17:33–49.

Ojima, D. S., T.G.F. Kittel, T. Rosswall, and B. H. Walker. 1991. Critical issues for understanding global change effects on terrestrial ecosystems. *Ecol. Applic.* 1:316–325.

Okkerman, P. C., E.J.V.D. Plassche, W. Sloof, C. J. VanLeeuwen, and J. H. Canton. 1991. Ecotoxicological effects assessment: a comparison of several extrapolation procedures. *Ecotox. Env. Safety* 21:182–193.

Parsons, K., A. Maccarone, and J. Brzorad. 1991. First breeding of double-crested cormorant (*Phalacrocorax Avritus*) in New Jersey. *Records NJ Birds* 17:51–52.

Peakall, D. B. 1992. *Animal Biomarkers as Pollution Indicators*. New York: Chapman and Hall.

SCOPE/MAB (Scientific Committee on Problems of the Environment/Man and the Biosphere Programme). 1987 (January). Definition and descriptiion of biosphere observation for studying global changes. SCOPE/MAB workshop, Paris, France.

Sheehan, P. J., D. R. Miller, G. C. Butler, and P. Bourdeau, eds. 1984. *Effects of Pollutants at the Ecosystem Level.* New York: Wiley.

Slovic, P. 1987. Perception of risk. *Science* 236:280–285.

Snowden, R. J., and I.K.E. Ekweozer. 1987. The impact of a minor spillage in the estuarine Niger delta. *Marine Poll. Bull.* 18:595–599.

Statistical Analysis Systems. 1985. *Statistical Analysis.* Cary, N.C.: Statistical Analysis Systems.

Suter, G. W. 1990. Endpoints for regional ecological risk assessments. *Environ. Manage.* 14:9–13.

Suter, G. W., L. W. Barnthouse, and R. V. O'Neill. 1987. Treatment of risk in environmental impact assessment. *Environ. Manage.* 11:295–303.

Vernberg, F. J., M. S. Guram, and A. Savory. 1977. Survival of larval and adult fiddler crabs exposed to AroclorR 1D16 and 1254 and different temperature–salinity combinations. In F. J. Vernberg, A. Calabrese, F. P. Thurberg, and W. B. Vernberg, eds., *Physiological responses of marine biota to pollutants,* pp. 37–50. New York: Academic Press.

Ward, D. V., and D. A. Busch. 1976. Effects of Tenefos, an organophosphorus insecticide, on survival and escape behavior of the marsh fiddler crab *Uca pugnax. Oikos* 27:331–335.

Ward, D. V., B. L. Howes, and D. F. Ludwig. 1976. Interactive effects of predation pressure and insecticide (Tenefos) toxicity on populations of the marsh fiddler crab *Uca pugnax. Marine Biology* 35:119–126.

11. Fish and Shrimp Populations in the Arthur Kill

John N. Brzorad and Joanna Burger

In 1989, the year before the Exxon oil spill in the Arthur Kill, many researchers were involved in studying the ecology of organisms living in the Arthur Kill region. Since we did not expect a large oil spill, we did not not tailor our research for a spill. None of us were prepared for the events that unfolded beginning January 1, 1990. Fortunately, we had 3 months of data from 1989 on faunal abundance, and this served to characterize the prespill shallow-water biota of the Arthur Kill region.

In field ecology it is often necessary to postulate effects without the help of rigorously controlled experiments. In the true sense of the controlled experiment, it is impossible to duplicate an ecosystem that would allow the comparison of copies of oiled and unoiled versions of the Arthur Kill. Ecologists are sometimes criticized for not having controlled all physical and biotic variables, making conclusions suspect. We argue that ecological studies, although not perfectly controlled, can nevertheless examine the effect of the disruption caused by oil spills. The goal in this chapter is to examine what impact the oil spills of 1990 had on shallow-water communities, particularly shallow-water fish and shrimp populations, in the Arthur Kill region.

Shallow-Water Fauna of the Arthur Kill

The Arthur Kill is commonly perceived by commuters passing by as a dead waterway. In fact, using various sampling methods, Howells and Brundage (1977) collected sixty-three species of fish, representing thirty-three families, in the area during surveys between 1972 and 1976. Winter flounder (*Pseudopleuronectes americanus*), hogchoker (*Trinectes maculatus*), bay anchovy (*Anchoa mitchilli*), and the mummichog (*Fundulus heteroclitus*) spawn in the Arthur Kill (Brundage 1977). Sixty species composing thirty-four families were found in Barnegat Bay, New Jersey, from a seine haul survey (Marcellus 1972). Although species counts are not directly comparable because of differences in sampling methods, the fish fauna of the Arthur Kill was found to be more diverse than expected. Because so many fishes occur in the study area, we focus on three species that are easily caught in beach seines and also are commonly found in less-polluted New Jersey and New York waters.

The most abundant year-round resident fish in the Arthur Kill is *Fundulus heteroclitus*, the mummichog (Brundage 1977; Howells and Brundage 1977; Brzorad and Burger 1990). Little is known about the winter habits of the mummichog; they may retreat into deep waters or bury into subtidal or intertidal muds (Chidester 1920). As the temperature rises in the spring, fish begin physiological preparations for breeding, which lasts from May to August. Mummichogs are most abundant from July to November, with young-of-the-year the most numerous year class during this period.

Because it is a year-round resident and can live up to 3 years in the polluted Arthur Kill (Toppin et al. 1987), the mummichog is more exposed to pollution than are other fish in the region. Larval fish hatched in the Arthur Kill will probably remain within the system because of very low flushing rates (Oey, Mellor, and Hires 1985). Lotrich (1975) found that mummichogs in a Delaware tidal creek have small home ranges, on the order of tens of meters, during the summer months.

Mummichog diets are diverse and vary with age of the fish. Young fish feed on benthic infauna and have been shown to reduce populations of polychaetes and mollusks (Kneib and Stiven 1982), but older fish prey on grass shrimp (*Palaeomonetes* spp.) in unpolluted areas. Mummichogs in the Arthur Kill and Rahway River feed extensively on decaying organic matter, or detritus (Brzorad and Burger 1990; Weis and Kahn 1991). The nutritive value of detritus in mummichog diets has been questioned: Prinslow et al. (1974) claim that detritus did not contribute to maintenance and growth in mummichogs when provided as a supplement, and Weisberg and Lotrich (1986) found that mummichogs ingest detritus when common food is depleted. This implies that mummichogs in the Arthur Kill and the Rahway River feed on detritus due to a lack of prey usually found in unpolluted marshes.

Unlike mummichogs, Atlantic silversides (*Menidia menidia*) are seasonal residents that arrive in the spring and depart for offshore waters in the autumn. Conover (1985) found that less than 1 percent of breeding populations are 2 years old. Pollution impacts would be most noticeable in this species if the large young-of-the-year population in July and August were affected. Winter pollution would show noticeable impacts only if its effects lasted into the summer months or occurred in offshore wintering areas. Silversides are mid-water foragers and live near the substrate and marsh grasses only during spawning (Conover and Kynard 1984). Large amounts of soluble petroleum (Boehm and Quinn 1973) have to be present in the water column to substantially impact this species.

The life cycle of the grass shrimp is brief and similar to that of silversides. Grass shrimp are predominantly annual, with less than 4.5 percent of breeding populations represented by 2 year olds (Welsh 1975). In the autumn, grass shrimp move into deep areas, and predation by mummichogs may significantly reduce their abundance, particularly in unpolluted areas. In a Rhode Island population, spawning occurred between May and July (Welsh 1975). Grass shrimp are important links between salt marsh and estuarine food webs (Welsh 1975). Salt marshes, dominated by cordgrass (*Spartina alterniflora*), are one of the world's most productive ecosystems (Mitsch and Gosselink 1986), but most of this productivity enters estuarine detrital food chains, with little going to terrestrial systems (Teal 1962). Grass shrimp chew on small cordgrass particles and create large amounts of surface area, which becomes inhabited by decomposer communities (Welsh 1975). Grass shrimp consume little macroalgae, but rather assimilate diatoms, green and blue-green algae, protozoans, rotifers, and nematodes (Morgan 1980), and these prey populations benefit from the reworking of cordgrass detritus. So these shrimp act as agents that make cordgrass biomass available to microbial food webs. In so doing, shrimp create conditions favorable for their own population. Shrimp also eat nematodes, anemones, ostracods, and the polychaete *Capitella capitata* (Kneib 1985), and as they forage they disturb the sediments, resuspending contaminants.

Effects of Pollution on Fishes

The water quality of the Arthur Kill has improved considerably since the 1960s, due in large part to the implementation of sewage treatment. Despite these improvements, because of concentrations of metals, chlorinated pesticides, PCBs, and polycyclic aromatic hydrocarbons in sediments, the Arthur Kill ranks among the top twenty polluted coastal and estuarine sites in the United States (National Oceanic and Atmospheric Administration 1988). Because of the Arthur Kill's history of pollution, other similarly polluted communities should be used as a reference in studying the impact of recent petroleum spills. A completely "clean" site is not a suitable control. The synergistic effects of many toxins together can cause more damage than the sum of the effects of single pollutants considered alone (Fingerman 1980). Thus, it is unwise to use unpolluted shallow-water communities as the only control in studying the effects of the 1990 oil spills.

Petroleum products are a complex mixture of aromatics; aliphatic;

and high-, middle-, and low-molecular-weight hydrocarbons (Clark 1989). Diesel fuel (no. 2 petroleum) contains high concentrations of aromatic hydrocarbons and is particularly toxic because of its carcinogenic qualities. In addition, biodegradation in the first 8 months after a spill reduces the straight-chain hydrocarbon fraction, leaving the aromatic fraction intact. So, on a volume basis, the toxicity of weathered no. 2 petroleum can increase before the aromatics are degraded (Blumer et al. 1970).

Because of petroleum's hydrophobic qualities, direct exposure of fishes to raw product alters cell membrane structure and function (Sanders et al. 1981). This can lead to abnormal gill function (Engelhardt, Wong, and Duey 1981) and physiological problems associated with altered oxygen absorption. The slimy, noncellular outer layer of fishes acts as a first line of defense against pathogens, so its disruption by petroleum may increase the occurrence of disease. For instance, the incidence of black-spot disease was positively correlated with indexes of habitat degradation along an urban gradient (Steedman 1991). Similarly, the incidence of fish tumors was higher in a polluted river than in an unpolluted control river (Brown et al. 1973).

Many fishes have the capability of detoxifying harmful compounds, but this incurs an energetic cost. Burns (1976) reports that liver enzyme systems (microsomal mixed-function oxidases [MFO]) become more active when mummichogs are exposed to petroleum. The levels of these enzymes are about twice as high in populations of fish living in petroleum-polluted waters than in fish from unpolluted waters. MFO systems require additional energy to function. Therefore, if an impacted population is to maintain its biomass over the course of a season, it must ingest more food than an unimpacted population requires. Petroleum contaminated food may then move through food webs rapidly as a result of heightened carbohydrate metabolism. Pink salmon (*Oncorhynchus gorbuscha*) fry fed oil-contaminated prey show lower growth rates then do control fry (Schwartz 1985). The growth decline may be a result of lower feeding rates but may also be due to elevated metabolic rates (Schwartz 1985). Not all individuals in a population forage in the same way or are as efficient (Ringler 1983, 1985; Ringler and Brodowski 1983), and those incapable of meeting their increased energy demands will eventually lose weight and perish. Mummichog populations in the Arthur Kill may be at a disadvantage in this regard. Weis and Khan (1991) found that fish collected from Piles Creek captured prey (juvenile guppies) less effectively than did fish from a cleaner site in Southampton, Long Island. At a given body length, the Arthur Kill fish weighed less than the Southampton fish. Weiss and Kahn (1991) suggest that poor prey-capturing abilities and diet explain the lower weight of Arthur Kill fish. This study may be complicated by the fact that fish from a northern and southern race were not distinguished (Able and Felley 1986).

Community Responses to Petroleum

Most research on petroleum toxicity examines the impact of oil on one species, usually in the laboratory setting. Few investigations are set in the field, and studies rarely examine how pollution alters entire communities. Such an approach is difficult because of the complexity of ecosystems under both stressed and unstressed conditions.

The unfortunate frequency and widespread impact of oil spills has, however, provided several instructive case studies. In March 1967, 150 kilometers of British coastline were damaged by the *Torrey Canyon* spill of Kuwaiti crude. Damage to the rocky intertidal communities was worsened because of treatment by chemical dispersants, and 9 to 10 years after the spill, heavily damaged areas had not fully recovered (Southward and Southward 1978). A large die-off of brown algae (*Fucus* spp.), grazers (limpets), and barnacles occurred in the oiled and treated areas. Green algae (*Enteromorpha* spp.) recolonized first, followed by *Fucus* spp. and later by limpets. Finally, barnacles returned, excluding limpets. Before complete recovery, the community experienced a decline in species richness (the total number of species) but an increase in biomass of a few early successional species (e.g., *Enteromorpha* spp.). This tendency is not limited to rocky intertidal zones. In subtidal soft-bottom habitats the opportunistic polychaete *Capitella capitata* invades areas denuded by petroleum spills and can remain in high numbers for up to a year after an accident, to the exclusion of most other species (Grassle and Grassle 1974; Sanders et al. 1981).

After the spill in the Persian Gulf, species richness declined and a biomass of tolerant infauna increased along a gradient of increasing sediment petroleum concentrations (Coles and McCain 1990). During recovery from the West Falmouth oil spill of 1969, variations in community composition were first reflective of successional changes related to repopulation but later became seasonal, as species composition approached the undisturbed state (Sanders et al. 1981). Recovery was not complete even after 5 years.

The present study examines the shallow-water fauna 1 year before and for 2 years after the Exxon Arthur Kill spill. The study was initiated in 1989 to understand prey dynamics of herons and egrets (Brzorad and Burger 1990). The most abundant shallow-water species—mummichogs, silversides, and grass shrimp—are eaten by both aquatic and terrestrial predators.

Mummichogs, Silversides, and Grass Shrimp

Methods

Two sampling sites in the vicinity of Pralls Island (Fig. 11.1) were chosen. The Rahway River site experiences lower salinities than does the Arthur Kill site located on Pralls Creek. Both sites are exposed to a 1.6-meter tidal range. Egrets and herons frequently forage at these sites.

A 30-foot bag seine (⅛-inch mesh size) was used to capture shallow-water fauna. The substrate in the sampling areas was deep mud, planks were laid in a square pattern on the mud surface so researchers could walk there without sinking. This ensured that the sampling effort was constant from one sample to the next, and that the sediments were not disturbed. Two people walked over the square plank pattern while pulling the seine during sampling (Fig. 11.2). A tide gauge was positioned below the spring tide low-water level at both sampling sites. Seining was always done at the same water level but at two different tide phases (ebb and flood).

Organisms in the seine were sorted and counted by species. Mummichogs and other fish were sifted through a fish sorter (see Fig. 11.2) into four size classes (A = largest, D = smallest), and individuals in each class were counted. We randomly chose subsamples of the D-size mummichogs, the silversides, and grass shrimp and recorded their total length. Wet weight was calculated from known length–weight relationships for mummichogs and silversides. Samples reported on here are from July through September in 1989, 1990, and 1991. To normalize the data, 1 was added to all species counts and the logarithm (base 10) was taken of this sum. A three-factor analysis of variance was used to analyze the data (PROC GLM, SAS release 5.2). It should be noted that the mummichog sorter did not divide the sample into discrete age classes but rather provided a quick way to characterize the size distribution of a given sample.

Results

Two primary changes in populations of fish and shrimp involved their abundance and size. The oldest, largest mummichogs (A-size) were more abundant in 1989 than in either 1990 or 1991, when no fish of this

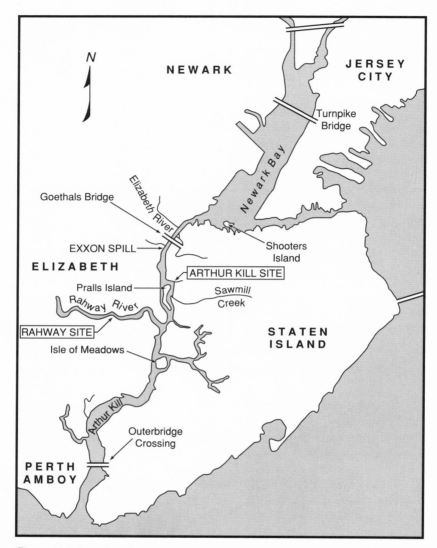

Figure 11.1. Location of the sampling sites in the Arthur Kill region.

size were caught (Fig. 11.3). More of these large fish were caught during ebb tide than during flood tide. The abundance of B-size mummichogs dropped sharply following the oil spill but climbed to its highest level in 1991 (Table 11.1). These fish were more abundant at the Arthur Kill site

Figure 11.2. The mummichog sorter.

Table 11.1. Differences in Mummichog Abundance by Size, Year, Site, and Tide

| | \multicolumn{8}{c}{Size} | | | | | | | |
	A $\overline{X}$	SE	B $\overline{X}$	SE	C $\overline{X}$	SE	D $\overline{X}$	SE
Year								
1989 (20)	1.3	0.5	6.9	1.7	30.6	9.2	546.0	149.1
1990 (26)	0	0	0.8	0.3	20.7	4.3	752.0	101.6
1991 (17)	0	0	12.3	3.7	132.0	38.1	2,325.0	632.7
Site								
Rahway (36)	0.4	0.2	5.3	1.8	65.0	20.4	1,575.6	313.6
Arthur Kill (27)	0.5	0.3	6.6	1.8	39.9	8.5	492.9	154.6
Tide								
Flood (31)	0.1	0.1	3.1	1.3	27.5	7.9	815.3	227.6
Ebb (32)	0.8	0.3	8.6	2.1	80.1	22.1	1,398.7	325.8

NOTE: $\overline{X}$ = average, SE = standard error. Sample size in parentheses.

than at the Rahway site, and more were caught during ebb tide than during flood tide. C-size mummichogs showed the same yearly trend as did the B-size fish; a postspill drop occurred in 1990, and the numbers rose in 1991 (Table 11.1).

At the Rahway site C-size mummichogs were more abundant during ebb tide than during flood tide over the study, but this difference did not occur at the Arthur Kill site (Fig. 11.4). The D-size mummichogs, which are the young-of-the-year, were the most abundant size class. Different trends are seen at the Rahway and Arthur Kill sites. At Rahway the young-of-the-year declined after the spill in 1990 and then reached their greatest abundance in 1991, as was true for the B- and C-size fish. The decline in 1990 was not evident for the flood-tide samples, and abundance rose steadily from 1989 to 1991. At the Arthur Kill site the numbers rose from 1989 to 1991, and this trend was strongest during ebb tide (Fig. 11.5). Young-of-the-year mummichogs were longer in 1990 than in either 1989 or 1991.

Silversides were more abundant at the Rahway site than at the Arthur Kill site, but the opposite was true for grass shrimp (Table 11.2). Grass shrimp declined after the oil spills in 1990 and had not yet fully recovered by 1991. The abundance of silversides increased from 1989 to 1991. Silversides showed a steady increase in size from 1989 through 1991.

Mummichog and silverside biomass rose each year from 1989 through

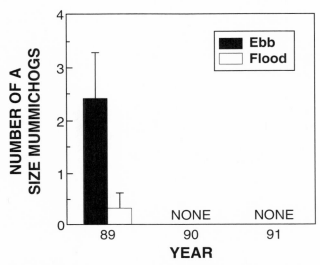

Figure 11.3. Changes in A-size mummichog abundance during ebb and flood tide from 1989 to 1991. Shown are means ± standard error

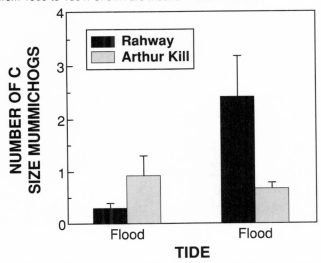

Figure 11.4. C-size mummichog abundance during ebb and flood tide at the Rahway and Arthur Kill sampling sites.

1990. More grams of mummichogs and silversides were caught at the Rahway site than at the Arthur Kill site. The weight of mummichogs was greater at ebb tide than at flood tide, but no significant difference existed for silversides (Table 11.3).

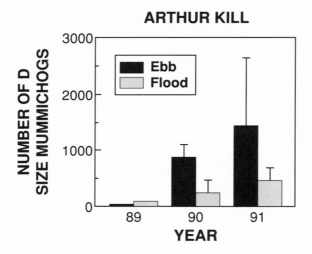

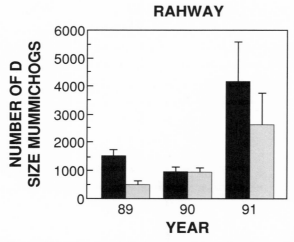

Figure 11.5. Changes in D-size mummichog abundance during ebb and flood tide at the Rahway and Arthur Kill sampling sites from 1989 to 1991.

The overall increase in the combined weight of mummichogs and silversides was greater at Rahway than at Arthur Kill. At Rahway the weight of mummichogs was greater at ebb tide then at flood tide, and much higher in 1991 then in either 1989 or 1990. A steady and less dramatic rise is seen at the Arthur Kill site (Fig. 11.6).

Table 11.2. Differences in Silverside and Grass Shrimp Abundance by Year, Site, and Tide

	Silversides		Grass shrimp	
	X̄	SE	X̄	SE
Year				
1989 (20)	66.4	18.5	445.0	179.8
1990 (26)	213.8	57.9	46.3	9.9
1991 (17)	644.2	207.1	283.3	160.2
Site				
Rahway (36)	404.2	96.5	118.6	32.2
Arthur Kill (27)	121.7	78.3	394.5	163.1
Tide				
Flood (31)	292.1	93.6	243.8	91.9
Ebb (32)	274.4	95.8	230.1	115.8

NOTES: X̄ = average, SE = standard error. Sample size in parentheses.

Table 11.3. Differences in Biomass (grams) of Mummichogs and Silversides by Year, Site, and Tide

	Mummichogs		Silversides	
	X̄	SE	X̄	SE
Year				
1989 (20)	843.4	197.6	46.7	25.1
1990 (26)	1,079.9	149.1	84.9	28.2
1991 (17)	3,041.7	641.8	614.7	282.3
Site				
Rahway (36)	2,046.8	346.4	320.5	138.1
Arthur Kill (27)	850.8	175.5	76.0	41.6
Tide				
Flood (31)	969.6	198.7	186.6	63.8
Ebb (32)	2,081.3	373.9	243.9	149.9

NOTE: X̄ = average, SE = standard error. Sample size in parentheses.

Discussion

A biological community is a dynamic composition of genetically unique species linked to one another by competition, predation, and symbiosis. Each species has different tolerances to disturbance. The damage done by petroleum accidents depends on the type and quantity of petroleum spilled, the time of year, and the type of community subjected to the toxic. The impact may be difficult to interpret. Some species, or age classes

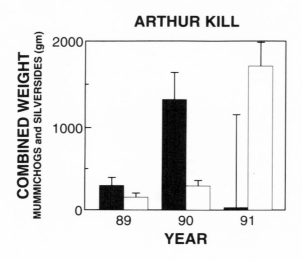

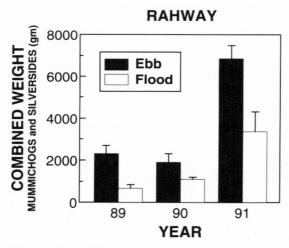

Figure 11.6. Combined weight changes of mummichogs and silversides during ebb and flood tide at the Rahway and Arthur Kill sampling sites from 1989 to 1991.

within a species, are killed outright; many others exhibit sublethal effects. Survivors must mobilize energy reserves necessary for detoxifying harmful petroleum components, many of which are carcinogenic. Still others may actually benefit from the disturbance caused by the spill. One of the complex effects of the 1990 spills on shallow-water communities in the Arthur Kill is shown in Figure 11.7.

In the autumn mummichogs from many regions of the Arthur Kill may

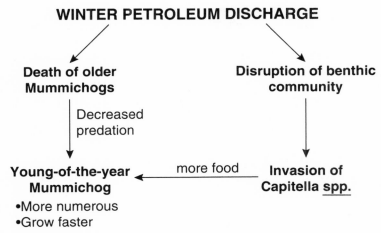

Figure 11.7. Model explaining sampling results in areas of petroleum discharge in the Arthur Kill.

converge into large wintering schools, and some of these schools probably wintered in areas impacted by oil. Generally, the autumn is a period of mass movement, and large schools of mummichogs may move into lower-salinity headwaters (Fritz, Meredith, and Lotrich 1975). In November 1988 a 3-meter-wide, 1-kilometer-long column of mummichogs was observed swimming upstream in the Rahway River (Brzorad, unpub. data). Small fish (3 to 5 cm total length) swam closest to shore in shallow water, while large fish (9 to 12 cm) swam in deeper water. This school probably contained just under one million fish. It is likely that many such large schools were moving during this period, attesting to the biomass accumulated during the growing season. The large-size classes were probably killed either at the time of the spill or soon after. In the late spring fish moved back into their summer home ranges. Thus, the similarity of the size structure between regions reflects the fact that they may have been exposed to petroleum while they were together in one large group during the winter. Some subpopulations such as those at the Arthur Kill site continued to live in degraded habitats in the summer, facing continued exposure to oil.

The Arthur Kill site was heavily oiled and experienced a large die-off of marsh cordgrass, whereas the Rahway site was only slightly oiled with no death of cordgrass. Changes in the abundance of small mummichogs reflect differences in summer conditions at the Rahway River and Arthur Kill sites, since at the time of sampling these young-of-the-year mummichogs had lived only in these areas. A large increase in *Capitella*, a pollution-resistant polychaete, was observed at the Arthur Kill site and served as a food source for young-of-the-year fish. No population explosion of *Capitella* was observed at the Rahway River site, and fish in this area could not supplement their diets with this polychaete.

A large increase in the number of the smaller-size mummichogs was observed the year after the spill at the Arthur Kill site. No such increase was seen at the Rahway River site, suggesting that the additional food at the Arthur Kill site improved chances of survival for the young-of-the-year. The additional food also influenced the growth rates of the smaller-size fish. Before the spill (1989) these fish grew steadily at both sites. In 1990, however, young-of-the-year at the Arthur Kill site were already considerably larger than young-of-the-year at the same site in 1989. There was a much smaller size difference at the Rahway River site between these 2 years (Fig. 11.8). This size class remained consistently larger at the Arthur Kill site but fluctuated at the Rahway River site. It is unclear why fish at the Arthur Kill site did not show additional growth in 1990.

Despite the abundance and growth enhancement of young-of-the-year fish at the Arthur Kill site, a population decline was observed for the largest-size classes of mummichogs at both sites. These fish had been in the intermediate-size class in the previous year, and their decline indicates poor overwintering conditions presumably worsened by petroleum exposure.

To evaluate the effect of winter conditions on mummichog survival it is possible to calculate a recruitment index for a size class of fish. In doing this three size classes of mummichogs were compared. The goal of calculating such an index was to examine winter survival of mummichogs. We hypothesize that conditions were less favorable in the 1989–90 winter than in the subsequent winter of 1990–91. Any adverse effect of cold-winter conditions would be augmented by petroleum toxicity. Since petroleum degrades with time and becomes covered with sediment, we suggested that overwintering conditions should have been better for mummichogs in 1990–91 than during the winter of the spill (1989–90).

Three size classes (B, C, and D) were examined here. No A-size fish were caught in 1990 or 1991. By growing, a fish in a given class is recruited to the next-larger class at some time the following year. Thus, a recruitment index was calculated by dividing the average number of fish in a size class by the average number of fish in the next-smaller size from the previous year (Table 11.4). There was poor recruitment (high mortality) of both C- to B-size, and D- to C-size classes in the winter of 1989–90 (.027 and .038, respectively). Recruitment improved, however, among the same size classes during the next winter (.58 and .17). This is

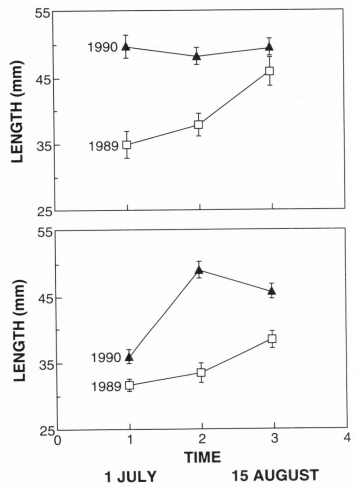

Figure 11.8. Length changes of young-of-the-year mummichogs at the Arthur Kill (top) and Rahway (bottom) sampling sites in 1989 and 1990.

consistent with our supposition that spilled oil becomes progressively less accessible with time, particularly in the Arthur Kill region, where high sedimentation rates exist.

Of course, an alternative hypothesis is that the winter of 1989–90 was colder than the winter of 1990–91 and that mummichogs died from exposure to cold conditions. Little is known about the effects of winter severity on the survival of mummichogs. Year-to-year catch per effort varied little for mummichogs between 1975 and 1978 in Barnegat Bay,

Table 11.4. Recruitment Index (RI) of Mummichogs to Various Size Classes in the Winters of 1989–90 and 1990–91

1989–90		RI
$\dfrac{\text{Avg. B-size mummichogs in 1990}}{\text{Avg. C-size mummichogs in 1989}} = \dfrac{0.84}{30.60} = 0.027$		
$\dfrac{\text{Avg. C-size mummichogs in 1990}}{\text{Avg. D-size mummichogs in 1989}} = \dfrac{20.96}{546.00} = 0.038$		

1990–91		RI
$\dfrac{\text{Avg. B-size mummichogs in 1991}}{\text{Avg. C-size mummichogs in 1990}} = \dfrac{12.35}{20.96} = 0.58$		
$\dfrac{\text{Avg. C-size mummichogs in 1991}}{\text{Avg. D-size mummichogs in 1990}} = \dfrac{132.00}{752.00} = 0.17$		

New Jersey (Tatham, Thomas, and Danila 1984), suggesting that winter severity may not affect survival. Additional years of study are necessary to elucidate this relationship.

Throughout the year of the spill (1990), oil remained near the sediment surface and contacted organisms that associate with the bottom. Sedimentation no doubt began to bury the oil throughout that year, so that by the next winter, less was accessible to fauna that associate with the sediments. This may explain the improved recruitment of mummichogs during the winter and early spring of 1991. Yet the absence of large mummichogs in 1991 may indicate that petroleum-induced mortality was still a factor.

Silverside numbers showed a dramatic increase between 1989 and 1991. As with grass shrimp, this annual species is present only in the warm months and may have benefited from the absence of large mummichogs in 1990. Mummichogs also eat silverside eggs (Conover 1985), so it is unclear why the population showed such a surge in 1991, when many B-size Mummichogs were also present. All trends induced by the oil spill must, however, be set in the context of natural fluctuations in age structure and changes in response to other abiotic factors such as the weather, particularly in winter.

Our interpretation of size-specific mortality and differential recruitment is consistent with the population trends of grass shrimp. Shrimp chew on cordgrass detritus lying on mud surfaces and probe into sediments to capture infauna. Shrimp would be particularly impacted when oil remains on the upper layer of these surfaces, as in 1990. The shrimp population dropped sharply in that year. This decline may have resulted from direct toxic effects of the oil, or the oil may have disturbed the microbial communities that live on the chewed detritus and are food for the shrimp. The rebound in shrimp numbers seen in 1991 may reflect diminished accessibility of the spilled oil. Population trends in this species are easier to interpret since only a small percentage of the population usually survives the winter. The size of the population in a given year reflects the current health of the habitat.

The differing responses of silversides and grass shrimp are instructive, since both species experience nearly complete population turnover in the course of 1 year. Therefore, any differences in response to oil cannot be due to differences in age structure but rather reflect microhabitat and behavioral differences exhibited by the species. Silversides are mid-water feeders and seldom come in contact with oil-laden sediments. Alternatively, grass shrimp are frequently associated with sediments and plant surfaces. Decline in grass shrimp abundance and an increase in silverside abundance are indicative of high petroleum loads in sediments and low petroleum loads in mid-water areas, respectively.

Conclusions

The oil spills in the winter and early spring of 1990 appeared to kill older mummichogs and thereby altered community structure. We suggest that the reduced predation by large mummichogs and ample food provided by the opportunistic pollution-resistant polychaete (*Capitella*) allowed the population of young mummichogs to increase. As a result, the age structure was shifted to the younger classes. Grass shrimp, being associated with sediments, also showed a decline after the spill but began to recover by the end of 1991. Population declines in grass shrimp may reduce the amount of detritus available to estuarine food webs. Silversides appeared to be unaffected by the oil spill: Their abundance steadily increased from 1989 to 1991.

We hope that this chapter can be helpful in any part of the world to prepare communities for industrial accidents. Accidents can occur anywhere, and citizens are limited in their effectiveness by the locations of these spills. It is unrealistic for the general population to get involved in monitoring oil spills on the open ocean, since few people venture to such places. Because the geography of coastal areas varies dramatically, it would be unwise for us to recommend a specific sampling method for all areas. Using beach seines is not recommended (and is dangerous) on wave-swept rocky shores or in mangrove thickets. Trawls cannot be used where boat navigation is difficult.

Sampling-gear limitations also dictate which biological communities are monitored. For example on a rocky, boulder-strewn coast it may be impossible to catch fish in an unbiased fashion given limited funds. Such places are more amenable for the study of encrusting biota such as barnacles, mussels, limpets, and algae.

Ecological principles apply wherever life is found. The processes of predation, competition, and symbiosis link species in a community. Food webs are also found in all communities, and it is possible to identify producers (plants), primary consumers (herbivores), secondary consumers (carnivores), and decomposers. These trophic levels are linked by predation and often by competition and symbiosis. Industrial accidents perturb these links, so the examination of as many trophic levels in a sampling program as possible is important, since changes in one level cascade to other levels. A basic question that should be asked by all those interested in monitoring communities is, What does this animal eat?

One of the first steps that citizens should take to be prepared for an industrial accident is to learn to identify the local flora and fauna of the area. The nature section of local bookstores should have a number of field guides covering plants, insects, fishes, reptiles and amphibians, birds, and mammals. Second, concerned citizens should initiate dialogue with local conservation groups, fishermen, hunters, farmers, and scientists who share an interest in the ecological welfare of a region. People from these groups will also be able to help in the identification of plants and animals.

In designing a sampling scheme it is important to consult with scientists with statistical knowledge so that a strong argument can be made from the data collected. Issues of gear bias should also be carefully thought through. No sampling method is perfect, and even with the most appropriate gear for a given habitat, some species or members of a species will avoid capture. It is ideal to use a number of different sampling methods, since a pollution effect argument can be strengthened if results from more than one type of gear are similar.

All communities could be concerned about industrial accidents, yet those in the vicinity of industrial plants should be particularly wary of possible mishaps. Our experience with the Exxon Arthur Kill spill indicated that issues of product transportation should be of great concern, since the oil was spilled while in the process of being moved through a pipeline. Communities should inquire about the exact location of pipelines and choose their sampling locations accordingly. The locations of tankers and barges should also be recorded as well as their destinations, names, and home ports. Data on the frequency of product transport should be in the public record and accessible to concerned citizens.

Sampling should begin *before* an accident occurs; this way if a pipe ruptures or a tanker runs aground data will already be available that characterizes the undisturbed ecology of the region (before–after method). Sampling should then continue in exactly the same manner as before the accident. A comparison can then be made of the biota before and after the accident.

If preaccident data are not available, then an alternative, albeit less favorable, approach is to monitor biota at varying distances from the accident (distance method). Strong arguments about ecological disruption can then be made if the ecosystem and its components can be shown to have been fairly uniform before the accident.

If resources are available, then a combination of the before–after and the distance methods should be used. Having data on the natural fluctuations in undisturbed areas during before-and-after time periods serves as an important control to the before–after data from the polluted site. This way fluctuations at the disturbed site can be distinguished from natural fluctuations.

Our best advice for concerned citizens is to be curious. Pursue the answers to your questions about the effects of oil by observing, reading, and querying scientists, naturalists, government officials, and leaders of industry. Keep a careful log of your questions, your observations, and the answers you discover.

Acknowledgments

This project would have been impossible without the help of many volunteers and field assistants. Deborah Begel, Susan Bissett, Gill Cardoso, Antonio Coppola, Cali and Dara Cutrone, Beverly DeAngelis, Wendy Dykes, Susan Elbin, Andrea Erikson, Ted Fog, Bob Gerwin, Amy Gross, Holly Hilton, Rafael Huaman, Jasmin Kilayko, Batina Lewan, Alan Maccarone, Lisa MacCollum, Kelly McFadden, Mary Anne McNulty, Neil Morganstein, Astrid Pujari, Greg Reviere, Danial Rosenblatt, Becky Thorne, Leo Tsao, and Cheryl Usher all helped catch fish and shrimp in the muddy waters of the Arthur Kill. Becky Thorne and Kelly Smith provided editorial help in the preparation of the manuscript. Richard McBride made many valuable suggestions on all parts of the manuscript.

The research was partially supported by the Hudson River Foundation under their Polgar and doctoral dissertation fellowships. Funds were provided by the states of New York and New Jersey as part of the damage assessment from the 1990 spill and by the Environmental and Occupational Health Sciences Institute (NIEHS grant ESO 5022). We would also like to thank the American Littoral Society for taking an interest in our project and providing the names of volunteers interested in learning about the fauna of the Arthur Kill.

Literature Cited

Able, K. W., and J. D. Felley. 1986. Geographical variation in *Fundulus heteroclitus:* Tests for concordance between egg and adult morphologies. *Amer. Zool.* 26:145–157.

Blumer, M., J. Sass, G. Souza, H. L. Sanders, J. F. Grassle, and G. R. Hampson. 1970. The West Falmouth oil spill: Persistence of the pollution eight months after the accident. Woods Hole Oceanographic Institution technical report no. 70-44.

Boehm, P. D., and J. G. Quinn. 1973. Solubilization of hydrocarbons by the dissolved organic matter in sea water. *Geochim Cosmochim. Acta* 37:2459–2477.

Brown, E. R., J. J. Hazdra, L. Keith, I. Greenspan, J.B.G. Kwapinski, and P. Beamer. 1973. Frequency of fish tumors found in a polluted watershed as compared to nonpolluted Canadian waters. *Cancer Res.* 33(2):189–198.

Brundage, H. M. III. 1977. Fish eggs and larvae of the Arthur Kill. *Underwater Naturalist.* 11(1):12–15.

Brzorad, J. N., and J. Burger. 1990. Heron and egret foraging behavior in the lower Hudson River estuary. Hudson River Foundation Polgar Fellowship report.

Burns, K. A. 1976. Microsomal mixed function oxidases in an estuarine fish, *Fundulus heteroclitus*, and their induction as a result of environmental contamination. *Comp. Biochem. Physiol.* 53B:443–446.

Burns, K. A., and J. M. Teal. 1979. The West Falmouth oil spill: Hydrocarbons in the salt marsh ecosystem. *Estuarine and Coastal Marine Science* 8:349–360.

Chidester, F. 1920. The behavior of *Fundulus heteroclitus* on the salt marshes of New Jersey. *Am. Nat.* 54:551–557.

Clark, R. B. 1989. *Marine Pollution.* Oxford: Clarendon Press.

Coles, S. L., and J. C. McCain. 1990. Environmental factors affecting benthic infaunal communities of the western Arabian Gulf *Mar. Envir. Res.* 29:289–315.

Conover, D. O. 1985. Field and laboratory assessment of patterns in fecundity of a multiple spawning fish: the Atlantic silverside, *Menidia menidia. Fish. Bull.* 83(3):331–341.

Conover, D. O., and B. E. Kynard. 1984. Field and laboratory observations of spawning periodicity and behavior of a northern population of the Atlantic silverside, *Menidia menidia. Environ. Biol. Fishes* 11:161–171.

Engelhardt, F. R. , M. P. Wong, and M. E. Duey. 1981. Hydromineral balance and gill morphology in rainbow trout *Salmo gairdneri*, acclimated to fresh and sea water as affected by petroleum exposure. *Aquat. Toxicol.* 1:175–186.

Fingerman, S. W. 1980. Differences in the effects of fuel oil, an oil dispersant, and three polychlorinated biphenyls on fin regeneration in the Gulf Coast killifish, *Fundulus grandis*. *Bull. Environ. Contamin. Toxicol.* 25:234–240.

Fritz, E. S., W. H. Meredith, and V. A. Lotrich. 1975. Fall and winter movements and activity level of the mummichog, *Fundulus heteroclitus*, in a tidal creek. *Ches. Sci.* 16(3):211–215.

Grassle, J. F., and J. P. Grassle. 1974. Opportunistic life histories and genetic systems in marine benthic polychaetes. *J. Mar. Res.* 32(2):253–284.

Howells, R. G., and H. M. Brundage III. 1977. Fishes of the Arthur Kill. *Proc. Staten Island Inst. Arts. and Sci.* 29(1):3–6.

Kneib, R. T. 1985. Predation and disturbance by grass shrimp, *Palaemonetes pugio*, in soft-substrate benthic invertebrate assemblages. *J. Exp. Mar. Biol. Ecol.* 93:91–102.

Kneib, R. T., and A. E. Stiven. 1982. Benthic invertebrate responses to size and density manipulations of the common mummichog, *Fundulus heteroclitus*, in an intertidal salt marsh. *Ecology* 63(5):1518–1532.

Lotrich, V. A. 1975. Summer home range and movements of *Fundulus heteroclitus* in a tidal creek. *Ecology* 56:191–198.

Marcellus, K. L. 1972. Fishes of Barnegat Bay, New Jersey, with particular reference to seasonal influences and the possible effects of thermal discharges. Ph.D. diss., Rutgers University, New Brunswick, N.J.

Mitsch, W. J., and J. G. Gosselink. 1986. *Wetlands*. New York: Van Nostrand Reinhold.

Morgan, M. D. 1980. Grazing and predation of the grass shrimp *Palaemonetes pugio*. *Limnol. Oceanogr.* 25(5):896–902.

National Oceanic and Atmospheric Administration. 1988. A summary of selected data on chemical contaminants in sediments collected during 1984, 1985, 1986 and 1987. National Technical Memorandum NOS OMA 44.

Oey, I. Y., G. L. Mellor, and R. L. Hires. 1985. Tidal modeling of the Hudson–Raritan estuary. *Estuarine, Coastal and Shelf Science* 20:511–527.

Penczak, T. 1985. Trophic ecology and fecundity of *Fundulus heteroclitus* in Chezzetcook Inlet, Nova Scotia. *Mar. Biol.* 89:235–243.

Prinslow, T. E., I. V. Valiela, and J. M. Teal. 1974. The effect of detritus and ration size on the growth of *Fundulus heteroclitus*. *J. exp. mar. biol. ecol.* 16:1–10.

Ringler, N. H. 1983. Variation in foraging tactics of fishes. In D.L.G. Noakes, ed., *Predators and Prey in Fishes*. The Hague: D. R. Junk.

———. 1985. Individual and temporal variation in prey switching by brown trout, *Salmo trutta*. *Copeia* 4:918–926.

Ringler, N. H., and D. Brodowski. 1983. Functional responses of brown trout, *Salmo trutta*, to invertebrate drift. *J. Freshwater Ecol.* 2(1):45–57.

Sanders, H. L., J. F. Grassle, and G. R. Hampson. 1972. *The West Falmouth oil spill: I. Biology*. Woods Hole Oceanographic Institution technical report 72-20.

Sanders, H. L., J. F. Grassle, J. R. Hampson, et al. 1981. *Long Term Effects of the Barge* Florida *Oil Spill*. Washington, D.C.: Environmental Protection Agency Office of Research and Development.

Schwartz, J. P. 1985. Effect of oil-contaminated prey on the feeding and growth rate of pink salmon fry (*Oncorhyncus gorbuscha*). In F. J. Vernberg, F. P. Thurberg, A. Calabrese, and W. Vernberg, eds., *Marine Pollution and Physiology: Recent Advances*, pp. 459–476. Belle W. Baruch Library in Marine Science, no. 13. Columbia: University of South Carolina Press.

Southward, A. J., and E. C. Southward. 1978. Recolonization of rocky shores in Cornwall after use of toxic dispersants to clean up the *Torrey Canyon* spill. *J. Fish. Res. Board Can.* 35(5):682–706.

Steedman, R. J. 1991. Occurrence and environmental correlates of black spot disease in stream fishes near Toronto, Ontario. *Trans. Amer. Fish. Soc.* 120:494–499.

Tatham, T. R., D. L. Thomas, and D. J. Danila. 1984. Fishes of Barnegat Bay. In M. J. Kennish and R. A. Lutz, eds., *Lecture Notes on Coastal and Estuarine Studies*, pp. 241–279. New York: Springer-Verlag.

Teal, J. M. 1962. Energy flow in the salt marsh ecosystem of Georgia. *Ecology* 43:614–624.

Toppin, S. V., M. Heber, J. S. Weis, and P. Weis. 1987. Changes in reproductive biology and life history of *Fundulus heteroclitus* in a polluted environment. In W. B. Vernberg, A. Calabrese, F. P. Thurberg, and F. J. Vernberg, eds., *Pollution Physiology of Estuarine Organisms*. pp. 177–178. Columbia: University of South Carolina Press.

Weis, J. S., and A. A. Khan. 1991. Reduction in prey capture ability and condition of mummichogs from a polluted habitat. *Trans. Amer. Fish. Soc.* 120:127–129.

Weisberg, S. B., and V. A. Lotrich. 1986. Food limitation of a Delaware salt marsh population of the mummichog, *Fundulus heteroclitus* (L.). *Oecologia* 68:168–173.

Welsh, B. L. 1975. The role of grass shrimp, *Palaemonetes pugio*, in a tidal marsh ecosystem. *Ecology* 56:513–530.

12. Gull and Waterfowl Populations in the Arthur Kill

Alan D. Maccarone and John N. Brzorad

The Exxon oil spill of January 1990 occurred during the winter, thus the migratory egrets, herons, and ibises that breed in the region were in distant wintering grounds and therefore were not immediately threatened. However, several thousand gulls and waterfowl overwinter each year in this area (NJDEP 1989), and many came into direct contact with this volatile petroleum. Hundreds of gulls and smaller numbers of waterfowl succumbed to the direct or indirect effects of this oil spill. Poisoning, starvation, drowning, predation, and hypothermia all contributed to the large number of carcasses found immediately after this and other 1990 spills (see Figs. 12.1 and 12.2).

This chapter discusses the potential effects of these oil spills on nesting-habitat selection by herring gulls (*Larus argentatus*) and of oil contamination on the reproductive biology and breeding success of gulls and waterfowl. It also summarizes the results of our ongoing study of nesting habitat selection by herring gulls on Pralls Island (Maccarone, Brzorad, and Parsons 1993); discusses reasons for the observed changes in the use of the Pralls Island shoreline by nesting gulls in the 2 years after the 1990 oil spills; and describes some of the potential short- and long-term effects of oil on diet and nest-site selection of gulls and other waterbirds.

Previous Research Findings

In 1986, we conducted an extensive census of the large gull population on Pralls Island. That year, volunteer workers for the Harbor Herons Project joined field biologists on a search over the entire 36-hectare (88-a.) island. Thereafter, only shoreline areas were censused for gull nests each year. In 1986, we noted that the west shore of Pralls Island was used by three to four times as many pairs of herring gulls than was the east shore (Fig. 12.3). Also, the great black-backed gull (*Larus marinus*), which comprises less than 5 percent of all breeding gulls on Pralls Island, was never found nesting along the east shore. Because of this substantial difference in the number of active herring gull nests between the west shore and east shore, and the absence of great black-backed gulls along the east shore, in 1987 we began to collect additional data

Figure 12.1. An oil-soaked great black-backed gull found floating near Isle of Meadows the day after the oil spill. The light no. 2 fuel oil may kill birds directly or else may cause hypothermia by destroying the insulative properties of the bird's feather coat.

Figure 12.2. The remains of a freshly killed and eaten gull, found on Isle of Meadows the day after the oil spill. Many large raptors were observed feeding on injured birds along the Arthur Kill during this period.

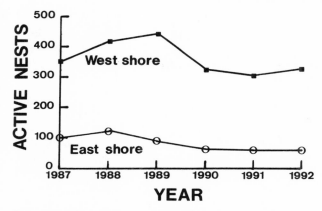

Figure 12.3. The number of active herring gull nests each year (1987–92) along two 1,000-meter-long sections of the west and east shores of Pralls Island. The west shore consistently contained three to four times as many nests as the east shore.

during our shoreline censuses to determine if a habitat-based explanation could be given for this difference.

Our preliminary field measurements from 1987 suggested that several abiotic factors affect shoreline nest density on Pralls Island. Based on our analyses of data collected during the next 2 years, we identified several major physical differences between the west shore and east shore. For example, the average width of the shoreline differs. The relatively narrow west shore has a mean width of 9.2 meters ± 6.3 meters (SD) (or 28.5 ft. ± 19.5 ft.), compared with the east shore, with a mean width of 14.8 meters ± 7.8 meters (or 45.9 ft. ± 24.2 ft.) (Maccarone et al. 1993). We then used multiple regression analysis to determine the factors that account for the differences in the density of herring gull nests between shores.

One important variable that accounted for differences in nest density was the slope of the shoreline to the water. The difference in slope was significant both in 1988 ($F = 6.11$, df = 1, 36, $p < .017$) and in 1989 ($F = 10.79$, df = 1, 37, $p < .003$). We found that the steep slope ($X = 5.9° ± 3.3°$) along the west shore protects that shoreline and prevents both occasional high spring tides as well as the constant waves from passing ships and tugboats from inundating gull nests. We believe that partly as a result of the steep slope, nest density along the west shore is higher than along the east shore. Also because of the steep slope, a large proportion of nests along the west shore are placed within 1 to 2 meters (3 to 6 ft.) of the water's edge. Nests placed close to the water's edge may enable young chicks to escape from potential predators by fleeing quickly into the water.

By contrast, along the east shore, where the average slope is only 1.5° ± 1.3° and many areas are virtually flat, nests are more vulnerable to high tides and wind-driven storm waves, which usually arrive from the east. Most nests along the east shore are placed far from the water's edge. Such nest placement, when combined with the tall, thick growth of cordgrass (*Spartina alterniflora*) along the east shore, may make escape into the water more difficult for young chicks.

A second variable associated with differences in herring gull nest density between the east shore and the west shore was the amount of wrack along the shoreline. Wrack included dead vegetation, automobile tires, lumber and other types of wood, and debris. Because Pralls Island is not maintained by any agency, this wrack has accumulated over many years and is removed only by the action of unusually high tides. The amount of wrack differed between the two shores in 1988, with the east shore containing more than the west shore. The amount of wrack did not differ between shorelines in 1989. The amount of shoreline wrack was important in explaining variability in nest density only in 1988 ($F = 7.48$, df $= 1, 36$, $p < .009$). These materials are often used by gulls to construct larger nests (Burger 1980) or to elevate their nests as protection against rising waters (Bongiorno 1970). Thus, the greater availability of wrack along the east shore may have enabled a greater number of gulls to nest along that shore than would otherwise have been possible. In fact, many such nests were placed inside automobile tires, atop large wooden planks, or within the mats of dead vegetation that had floated ashore.

A third important variable was the amount of sand and gravel (as opposed to vegetation) along the shore. Sand and gravel are preferred nesting substrates for herring gulls (Pierotti 1982), and such areas on Pralls Island often had the highest nesting densities. Sandy areas devoid of growing vegetation may facilitate landing and takeoff, as well as visual displays between nesting pairs. More important, open areas may allow nestlings unimpeded access to the water to escape danger. It is noteworthy that virtually no areas of sand or gravel were found along the east shore, which might also help to explain why that shore is so sparsely settled by herring gulls.

Shoreline measurements, including slope, substrate type, and shoreline width, were taken in 1988, 1989, and again in 1990, the year of the oil spills. After 1990, we discontinued measuring shoreline characteristics because we were satisfied with the consistency from one year to the next and the ability of these three variables to account for observed differences in the number of gull nests along the two shores of Pralls Island.

When the series of oil spills began early in 1990, we had already collected data on herring gull breeding populations and habitat selection during 3 consecutive years (1987–89) and had taken measurements of shoreline characteristics for 2 of those years (1988 and 1989). Based on our analysis of these data, we could support statistically an explanation of why one shore was used consistently by more pairs of herring gulls than was the other shore. Figure 12.3 shows the number of active nests located along the east shore and west shore of Pralls Island from 1987 to 1992. These baseline data enabled us not only to document changes in the gulls' nesting patterns in 1990, the breeding season immediately following the oil spills, but also to measure later responses by gulls in 1991 and 1992.

Causes of Oil Contamination of Gulls and Waterfowl

There are many ways in which waterbirds can come into direct or indirect contact with oil spills, such as the ones that occurred in the Arthur Kill and nearby waterways during the first half of 1990. Here we examine how the patterns of nest-site selection by gulls on Pralls Island, as described in the preceeding section, might affect their reproductive success after such oil spills. We also examine how the foraging behavior and diets of gulls, waterfowl, and other aquatic birds might expose both adults and their offspring to the lethal and sublethal effects of discharged petroleum products. In the last section, we review the current literature on the physiological response by gulls and waterfowl to contamination by various petroleum products.

On Pralls Island, and on the other islands in the Arthur Kill, many gulls build their nests close to the water, while other pairs construct their nests farther inland. The greatest danger from oil faces gulls whose nests are close to the water, because these birds may come into direct contact with the oil as it washes ashore. Even those nests placed farther from the water's edge, or inland, may be lined with materials that have come into contact with oil. Those herring gulls that nest along the nearly flat east shore of Pralls Island or similar flood-prone areas faced the most severe threat of contamination. Although the east shore has been rather sparsely colonized by nesting gulls throughout the 6 years of our study, many of these nests are unprotected from high waters and wind-driven spray. Flood tides are an effective method of delivering oil directly to eggs and chicks.

The dense shoreline vegetation on both sides of the island (mainly *Spartina alterniflora* and *S. patens*) was covered and repeatedly immersed in oily water in 1990. Both adults and ambulatory chicks

Figure 12.4. Shoreline vegetation on Pralls Island killed after prolonged exposure to water contaminated by no. 2 fuel oil. The usually lush carpet of *Spartina alterniflora* contributes nutrients for the estuary and provides cover and nesting material for many breeding birds.

frequently came into direct contact with this oiled vegetation, and their feathers may have become contaminated. When no. 2 oil was applied experimentally to the breast feathers of adult gulls, who were then allowed to transfer the oil to their eggs, embryo mortality exceeded 40 percent, compared with 2 percent of control eggs (Kings and Lefever 1979).

Frequent preening by waterbirds transfers the oil from the plumage into the bird's digestive or respiratory tract, often with lethal results (Eastin and Hoffman 1979). Even the wrack, which is used by many gulls to elevate their nests above the water, may have drifted in the oily waters before finding landfall on Pralls Island or other nearby gull breeding islands.

Finally, through prolonged contact with the oiled waters, some of the *Spartina*, which usually grows along both shorelines, was destroyed in 1990. By 1992, the most recent year of our investigation, it had yet to recover fully (but see Chapter 8). The damage to vegetation was primarily along the east shore of Pralls Island (Fig. 12-4). The lush carpet of *Spartina* helps to conceal sparsely placed nests and their contents from potential predators. With its destruction from contact with the oil, eggs

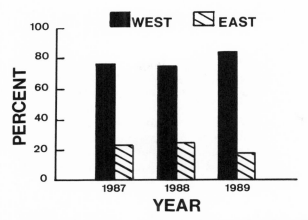

Figure 12.5. The percentage of active herring gull nests each year along two 1000-meter-long sections of the west shore and east shore of Pralls Island. Nests along the west shore decreased for the first time in 1990, the year of the oil spills; those along the east shore continued to decline from the previous year.

and chicks were more exposed to aerial predators such as great black-backed gulls and crows (*Corvus brachyrhynchos* and *C. ossifragus*).

The percentage of shoreline gull nests on Pralls Island between 1987 and 1989, the first 3 years of our study, are shown in Figure 12.5. The number of active nests built along the west shore increased steadily over the 3-year period preceding the oil spills. This increase was then followed by a sharp drop in the number of active shoreline nests in 1990, and then by another, smaller decrease in 1991. Only in 1992 did the number of nests along the west shore again begin to increase.

Changes in the size of an established breeding population may occur through various causes; thus, we are unable conclusively to ascribe this 2-year decrease in the number of herring gull nests along the west shore specifically to the oil. The decline in the number of nests may have resulted from disturbance in the form of increased human activity for cleanup and monitoring efforts or to the intensified research activity after the oil spills. Even in a highly urbanized area such as the Arthur Kill, this disturbance may have been sufficient to cause many pairs of herring gulls to nest elsewhere. A similar phenomenon was noted by Mueller and Glass (1988) in Texas: Oil-drilling activities during the spring breeding season caused many laughing gulls (*Larus atricilla*) and Forster's terns (*Sterna forsteri*) to relocate their nests. When this activity was terminated 2 years later, the number of nests of both species increased. On Pralls Island, the number of west shore nests increased in 1992, 2 years after the first oil spills and associated activities occurred. Finally, because we did not measure adult mortality, egg viability, or offspring survival of herring gulls either before or after these oil spills, we cannot determine whether the oil spills affected reproductive success.

Because the east shore historically has contained many fewer nests than the west shore, the fluctuations along that shore appear less pronounced. The most recent decline in the number of herring gull nests along the east shore follows a trend already begun the year before the spills. Thus, it is difficult to ascribe this continued decrease only to the oil spills and their aftermath and not to some combination of factors that might include the oil spills and some ecological factors or some other form of disturbance. It is likely that several factors account for nest-site selection and nest density along this shore (Maccarone, Brzorad, and Parsons 1993). Unlike with the west shore, however, the number of nests placed along the east shore did not increase in 1992, 2 years after the oil spills. The oil spills, then, may have have merely accelerated an existing shift in nest-site selection by these herring gulls.

Even in 1992, many gulls along the steeper, more well protected west shore still fashioned nests from the oil-soaked materials found there. This problem of using contaminated nesting material may also affect the few (six to ten) pairs of Canada geese that breed each year on Pralls Island. The geese usually build nests close to the water's edge and may face the same dangers from direct contact with the water as do many of the gulls. Even in the absence of large oil spills such as those in 1990, oil contamination is a constant threat to waterbirds in this region.

The dangers of direct contact with oil also affect other waterbirds nesting in this area, such as rails, long-legged wading birds, and the small population of double-crested cormorants (*Phalacrocorax auritus*) that recently began nesting on an abandoned dry dock at nearby Shooters Island (Parsons, Maccarone, and Brzorad 1991). Besides Canada geese, gadwalls (*Anas strepera*) and mallards (*A. platyrhynchos*) are the most common species of waterfowl breeding in this area. In contrast to the geese, these species of ducks build their nests inland and are less likely to come into direct contact with oil.

Regardless of location, proximity to water, or the material used to build and line nests, these waterbirds obtain food for themselves and their offspring from the Arthur Kill and surrounding waters, all of which were affected to some degree by the many oil spills. The ducks and geese, both primarily herbivorous, feed at a lower trophic level than the other species of waterbirds and might not suffer the damaging effects of biomagnification to the same degree (Cunningham and Saigo 1992). However, they still may have encountered lower food availability, owing to the localized destruction of aquatic vegetation. In addition, many of these waterfowl foraged on living but oiled vegetation, and their wet feathers may have contaminated their nests and eggs when they returned from feeding. Finally, the aquatic invertebrates that constitute a portion of their diet, especially during the breeding season, may have accumulated some of the toxins from the water or sediment.

The dangers associated with the consumption of contaminated prey may also face the hundreds of glossy ibises (*Plegadis falcinellus*) that forage for food in the mud flats and shallows of the Arthur Kill and its tributaries (see Chapter 13). By contrast, the cormorants are strictly piscivorous, as are several species of the wading birds that breed on these islands, and fish taken from local waters may have accumulated high concentrations of petroleum products, as well as the heavy metals and other contaminants found in these waters.

Although the name "herring gull" suggests that these birds subsist on fish, herring gulls are often scavengers (Tinbergen 1960), and our experience indicates that the gulls obtain a large portion of their diet from the nearby Fresh Kills landfill, located on Staten Island. Nevertheless, herring gulls and great black-backed gulls often take fish from the Arthur Kill and other nearby waterways, which may be contaminated by the oil.

In summary, contact with oil may result in: (1) coating of the adult plumage as a result of swimming, feeding, or other activities; (2) the ingestion of oil as a result of preening or eating contaminated food items; (3) the feeding of contaminated food items to nestlings; and (4) the transfer of oil from adult plumage to the nest, eggs, or nestlings.

Effects of Petroleum Contamination on Gulls and Waterfowl

A large and growing body of literature, both experimental and field-based, describes the effects of oil and other petroleum products on waterbirds. Many of the field studies were carried out as a result of large oil spills, which always seem to affect sensitive estuaries and beaches. Such oil spills have included those off the coast of Brittany, France, caused by the breakup of the *Amoco Cadiz;* one off the coast of California at Santa Barbara, caused by a tanker fire; the grounding of the *Exxon Valdez* in Prince William Sound, Alaska; and the oil spill at Firth of Forth, Great Britain. In the wake of the huge 1989 *Exxon Valdez* spill in Alaska, an estimated one hundred thousand to three hundred thousand seabirds were killed, especially alcids, as were thousands of gulls and waterfowl (Piatt et al. 1990). In the Firth of Forth incident in 1978, large numbers of grebes and ducks were either killed or incapacitated; some oiled birds managed to fly inland before they died or were rendered flightless by their coating of oil (Campbell, Standring, and Cadbury 1978).

Although large oil spills such as those just mentioned focus local and often worldwide attention on the problem of water pollution, only about 35 percent of all the petroleum released into waters is a result of such massive spills (Biderman and Drury 1980). The remainder finds its way into the oceans, bays, and estuaries of the world from countless smaller discharges and represents a chronic presence and a constant threat to wildlife.

Contact with and contamination by oil and other petroleum products affect birds in several ways. The extent to which the oil affects a bird differs according to the species of bird, the time of the bird's life stage when this contact occurs, the type of petroleum product involved, the amount of time between the release of the oil and its contact with the bird (the degree of "weathering" of that oil), and the length of contact with that oil (King and Lefever 1979; Biderman and Drury 1980; Peakall, Jeffrey, and Miller 1985).

In both experimental and field studies of contamination with oil, several species of birds may be examined or tested simultaneously. Often there are interspecific differences in the severity of damage or level of mortality. For example, Peakall et al. (1987) speculate that pursuit-diving seabirds, such as many ducks and alcids, are not as severely affected by a large oil spill as are many surface-feeding birds, such as gulls. This is because the oil, which is often mixed with dispersants, remains close to the surface. Thus birds that spend all their time near the surface have high exposure to the oil.

Interspecific differences in the effects of contact with oil have been found at the developmental stage as well. For example, Eastin and Hoffman (1979) found differential embryo mortality among different species of waterbirds subjected to the same regime of experimentally oiling the eggshells. These included six species in three orders: three larids (laughing gulls, great black-backed gulls, and sandwich terns, *Sterna sandvicensis*), two ducks (mallards and eiders, *Somateria mollisma*), and an ardeid (the tricolored heron, *Hydranassa tricolor*). As with many other mutagens and teratogens, the effects of petroleum contamination are more pronounced at earlier than later stages of avian embryological development. During this critical period, deformities and mortality attributable to oil are more common than during later stages of development (Biderman and Drudy 1980; Lewis and Malecki 1983).

These and other experimental studies with no. 2 fuel oil, the type of oil that leaked from the Exxon pipeline into the Arthur Kill, suggest that amounts as small as 5 microliters (.003 oz.) may cause high mortality in some species of waterbirds. As little as 50 microliters (.003 oz.) of oil applied to an egg can cause complete mortality in mallards (Eastin and

Hoffman 1979). In general, the longer the oil has been in contact with air and water and has "weathered," the less severe will be the damage to those birds exposed to its effects. However, even weathered oil can cause embryological damage to some birds, including gulls and herons (Macko and King 1980).

Beyond the actual mortality caused to adults and young, obviously the most severe effect of contact with petroleum products, oil contamination may also retard the maturation of ova in females, which might reduce future reproductive efforts (Hoffman and Albers 1984; Szaro 1977). The types of nonlethal damage caused to embryos and young birds include improper bone growth and bone deformities, bill abnormalities, stunted growth, liver and kidney damage, slow or abnormal behavioral responses, and the failure of flight feathers to grow properly (Biderman and Drury 1980). Any combination of these effects would restrict post embryonic development and inhibit a young bird's ability to survive and fledge. It is difficult to generalize about the potential dangers to aquatic birds from contact with oil, as these dangers differ greatly from one species to the next.

Type of diet and foraging method also may determine the severity of the effects of oil on waterbirds. For example, food gathering may be by surface feeding, as with many gulls and skimmers; by plunge diving for fish or squid below the surface, as with terns and gannets; or by sifting through mud and other sediments for small invertebrates, as with ibises. This latter type of feeding method may be especially dangerous, because small sediment-dwelling invertebrates are exposed for long periods to the toxins trapped there, and their tissues may contain high levels of these toxins (Cunningham and Saigo 1992). Other feeding methods by aquatic birds include capturing small fish from shallows, as with egrets and herons, and consumption of aquatic vegetation, as with waterfowl. Food contaminated by oil but still eaten by birds represents a decrease in food quality. However, other prey species or food organisms killed by contact with the oil, such as aquatic vegetation, small invertebrates, and fishes, represent a decrease in food availability. Waterbirds and the estuarine food webs of which they are a part may easily be disturbed when such changes occur at lower trophic levels.

A second factor that may account for interspecific differences in a bird's susceptibility to an oil spill involves aspects of the breeding biology of the species. For example, on Pralls Island and other local islands, both shore-nesting species (many pairs of gulls and the Canada geese) and species that build nests inland, farther from direct contact with the waterborne oil, are found. The latter category includes the many pairs of herring gulls and most great black-backed gulls, as well as all ducks that breed on this island. Habitat selection is therefore very important in determining the extent to which eggs and nestlings come into contact with oil. Also, the materials used to construct and line the nest and the places from which these materials are collected are elements of breeding biology that affect the degree and duration of exposure to oil.

While large oil spills are dramatic and galvanize public attention on the problems of pollution, it is instructive to remember that if every major oil spill were somehow prevented, at least two-thirds of the present amount of discharged oil would still find its way into the water through smaller, chronic releases.

Conclusions

Several accidental discharges in early 1990 released 4 million liters of petroleum products into the already-polluted Arthur Kill waterway. This estuary is used by overwintering waterbirds and is an important breeding area for thousands of gulls, waterfowl, and long-legged wading birds. This chapter documents changes in the size of the breeding population and nest-site selection by gulls on one of the islands affected by the petroleum discharges. Also discussed are the following: (1) the potential effects of these oil spills on the reproductive biology and breeding success of gulls and waterfowl, (2) how patterns of nest-site selection by gulls might affect their reproductive success after such oil spills, and (3) how the foraging methods and diets of gulls, waterfowl, and other aquatic birds might expose adults and offspring to the lethal and sublethal effects of petroleum products.

The study produced several major conclusions: (1) The dense shoreline vegetation was covered and repeatedly immersed in oily water, exposing adults and chicks to oil and possibly contaminating their feathers. Frequent preening by waterbirds may transfer the oil from their plumage into their digestive or respiratory tract. Even the wrack, which is used by many gulls to elevate their nests above the water, may have drifted in the oily waters before drifting ashore. (2) The oil may also affect other waterbirds nesting in this area, such as rails, long-legged wading birds, waterfowl, and cormorants. These waterbirds obtain food for themselves and their offspring from the Arthur Kill. Small fish, aquatic invertebrates, and aquatic vegetation may have accumulated some of the toxins from the water or sediment. (3) Oil-caused damage to the usually dense shoreline vegetation exposed gull nests and their contents to potential predators. (4) The severity of the effects of petroleum contamination on waterbirds varies by species and reflects different degrees of contact with the oil, the type of diet and foraging method used by each species, and various aspects of the breeding biology of the different species.

Acknowledgments

We thank Katharine C. Parsons and the volunteers of the Harbor Herons Project for their assistance in conducting censuses of gulls and for helping to take measurements of Pralls Island shoreline characteristics. We especially would like to thank Mary Kearns-Kaplan and Carolyn Summers for their abundant help in its many forms. The City of Elizabeth (New Jersey) Police Department provided a ride to Pralls Island on January 2, 1990, the day of the Exxon spill. This enabled us to locate and rescue several Canada geese that otherwisse might have perished. John Papetti and the Elizabeth Marina provided accommodations and cooperation during our entire study, for which we are extremely grateful.

Literature Cited

Biderman, J. O., and W. H. Drury. 1980. *The Effects of Low Levels of Oil on Aquatic Birds.* U. S. Fish and Wildlife Service, Biological Service Program.

Bongiorno, S. F. 1970. Nest-site selection in adult laughing gulls (*Larus atricilla*). *Animal Behaviour* 18:434–444.

Burger, J. 1980. Nesting adaptations of herring gulls to salt marshes. *Biology of Behaviour* 5:147–162.

Campbell, L. H., K. Standring, and C. J. Cadbury. 1978. Firth of Forth oil pollution incident, February 1978. *Marine Pollution Bulletin* 9:335–339.

Cunningham, W. P., and B. W. Saigo. 1992. *Environmental Science.* New York: William C. Brown.

Eastin, W. C., and D. J. Hoffman. 1979. Biological effects of petroleum on aquatic birds. In C. C. Bates, ed., *Proceedings of the Conference on Assessments of Ecological Impacts of Oil Spills*, pp. 561–582. Arlington, Va.: American Institute of Biological Sciences.

Hoffman, D. J., and P. H. Albers. 1984. Evaluation of potential embryotoxicity and teratogenicity of 42 herbicides, insecticides, and petroleum contaminants to mallard eggs. *Archives of Environmental Contamination and Toxicology* 13:15–17.

King, K. A., and C. A. Lefever. 1979. Effects of oil transferred from incubating gulls to their eggs. *Marine Pollution Bulletin* 10:319–321.

Lewis, S. J., and R. A. Malecki. 1983. Reproductive success of great black-backed and herring gulls in response to egg oiling. In: D. Rosie and S. N. Barnes, eds., *The Effects of Oil on Birds: Physiological Research, Clinical*

Applications, and Rehabilitation, pp. 98–112. Wilmington, Del.: Tri-state Bird Rescue & Research.

Maccarone, A. D., J. Brzorad, and K. C. Parsons. 1993. Factors affecting habitat selection in a population of herring gulls nesting in an urban estuary. *Colonial Waterbirds* 16:216–220.

Macko, S. A., and S. M. King. 1980. Weathered oil: Effect on hatchability of heron and gull eggs. *Bulletin of Environmental Contamination and Toxicology* 25:316–320.

Mueller, A. J., and P. O. Glass. 1988. Disturbance tolerance in a Texas USA waterbird colony. *Colonial Waterbirds* 11:119–122.

New Jersey Department of Environmental Protection (NJDEP). 1989. *Aerial Waterfowl Surveys*. Trenton: New Jersey Department of Environmental Protection.

Parsons, K. C., A. D. Maccarone, and J. Brzorad. 1991. First breeding record of the double-crested cormorant (*Phalacocorax auritus*) in New Jersey. *Records of New Jersey Birds* 17:51–53.

Peakall, D. B., J. A. Jeffrey, and D. S. Miller. 1985. Weight loss of herring gulls exposed to oil and oil emulsion. *AMBIO* 14:108–110.

Peakall, D. B., P. G. Wells, and D. Mackay. 1987. A hazard assessment of chemically dispersed oil spills and seabirds. *Marine Environment Research* 22:91–106.

Piatt, J. F., C. J. Lensink, W. Butler, M. Kendziorek, and D. R. Nysewander. 1990. Immediate impact of the *Exxon Valdez* oil spill on marine birds. *Auk* 107:387–397.

Pierotti, R. 1982. Habitat selection and its effects on reproductive output in the herring gull in Newfoundland. *Ecology* 63:854–868.

Szaro, R. C. 1977. Effects of petroleum on birds. *Transactions of the Forty-second North American Wildlife and Natural Resources Conference:* 374–381.

Tinbergen, N. 1960. *The Herring Gull's World*. London: Collins Press.

(King and Lafever 1979). Very small amounts of oil result in significantly lower hatching success and malformed surviving hatchlings. Deformities include brain, bill, and eye defects (Hoffman and Albers 1984). Chicks hatching from oiled eggs often show depressed growth rates, delayed fledging, and poor survival (Miller, Peakall, and Kinter 1978; Boersma 1988; Butler 1988). Several studies have implicated polycyclic aromatic hydrocarbons (PAHs) as the most toxic compounds in petroleum (Peakall, Miller, and Kinter 1983; Hoffman and Albers 1984). They are therefore the components most likely to produce the aforementioned effects.

Because of the complexity of petroleum mixtures, tissue concentrations are difficult to obtain and interpret (Szaro 1977; Ohlendorf, Klaas, and Kaiser 1978). Some studies show selective retention of branched, cyclic, and aromatic hydrocarbons in higher organisms, suggesting biomagnification (Ohlendorf, Klaas, and Kaiser 1978; Stickel and Dieter 1979).

Few long-term studies have been conducted on bird populations impacted by oil. Colony site fidelity and breeding success in years following an oil spill were abnormally low in a population of wedge-tailed shearwaters (*Puffinus pacificus*) (Fry et al. 1986) but not in a population of Leach's storm petrels (*Oceanodroma leucorhoa*) (Butler 1988).

The objective of this chapter is to examine the effects of oil contamination on the feeding and breeding ecology of the major groups of aquatic bird species utilizing the kills.

Seabirds

Cormorants

Double-crested cormorants (*Phalacrocorax auritus*) have undergone dramatic population fluctuations in the northeastern United States since European settlement in the seventeenth century (Hatch 1984). Fisherman believed cormorants competed with them for food, and so the cormorants were killed. Following local extirpation, remnant cormorant populations from Canada reclaimed historical breeding sites along the New England coast from 1920 to 1950. Continuing south, cormorants colonized coastal New York by 1977 (Buckley and Buckley 1984) and bred initially in the kills in 1987 (Parsons, Maccarone, and Brzorad 1991).

Cormorants are present year-round in the estuary, and four individuals were found dead by wildlife search-and-rescue teams during the 6 weeks following the Exxon oil spill (Table 13.1). Although the immediate impact of 1990's oil spills on cormorants appears minimal, population growth at the newly established Shooters Island cormorant colony in the Kill van Kull was less than expected in 1990 ($p < .05$). By 1991 double-crested

Table 13.1. Dead Water Birds Recovered Following the Exxon Bayway Oil Spill (Jan. 3 to Feb. 12, 1990) in the Arthur Kill/Kill van Kull Area of New York Harbor

	No. of dead individuals
Seabirds	
Great black-backed gull (*Larus marinus*)	103
Herring gull (*Larus argentatus*)	103
Ring-billed gull (*Larus delawarensis*)	9
Bonaparte's gull (*Larus philadelphia*)	2
Unknown gull	72
Double-crested cormorant (*Phalacrocorax auritus*)	4
Waders	
Black-crowned night heron (*Nycticorax nycticorax*)	14
Green-backed heron (*Butorides striatus*)	3
Unknown heron	2
Waterfowl	
Mallard (*Anas platyrhynchos*)	135
American black duck (*Anas rubripes*)	76
Gadwall (*Anas strepera*)	30
Canada goose (*Branta canadensis*)	16
Canvasback (*Aythya valisineria*)	13
Bufflehead (*Bucephala albeola*)	8
Scaup (*Aythya* spp.)	5

SOURCE: Louis Berger & Associates, Inc., 1991.

cormorants had resumed their prespill average population growth rate of eleven pairs per year (Fig. 13.1).

Gulls

Gulls are ubiquitous members of coastal ecosystems and are highly abundant in New York Harbor year-round. Prominent in the estuary are herring gulls (*Larus argentatus*) and great black-backed gulls (*L. marinus*).

Wintering gulls in urban areas make use of artificial food sources, such as landfills, which characterize urban landscapes and augment or replace the gulls' natural diet of aquatic organisms (Monaghan 1979; Hackl and Burger 1988). Their ability to exploit non-natural food resources and the protection awarded them through federal and state laws have resulted in rapid population growth for many species in the northeastern United States over the past half century (Kadlec and Drury 1968).

As populations have increased, modern breeding ranges have expanded southward from Canadian strongholds. Although herring gull populations stabilized in coastal New York over the period 1985 to 1989, great black-backed gulls were still increasing prior to 1990's oil spills (Downer and Liebelt 1990).

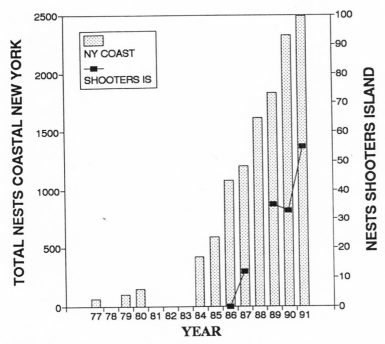

Figure 13.1. Abundance of nesting double-crested cormorants in coastal New York and Shooters Island, 1977–91.

A total of 289 dead gulls were retrieved from the Arthur Kill during the 6 weeks following the Exxon spill (Table 13.1). More gulls than any other bird group were killed in the aftermath of the spill. Gull recoveries peaked approximately 2 weeks after the spill, but consistent losses were documented until the search-and-rescue operation was discontinued on February 12, 1990 (Fig. 13.2).

Abundance of nesting herring gulls in the metropolitan area of New York remained fairly stable at approximately five thousand pairs following the oil spills of 1990 (data from Jamaica Bay were not collected in 1991) (Fig. 13.3A and B). In contrast, New York Harbor's great black-backed gull population dropped precipitously beginning in 1990 (Fig. 13.3C and D). Great black-backed gulls nesting in New York Harbor (including the East River, Upper and Lower New York bays, Newark Bay, and the kills) consistently outnumbered populations in Jamaica Bay and Hempstead by 2 to 1. In 1990, an increase in nesting great black-backs in Jamaica Bay and Hempstead accounted for gulls lost from New York Harbor.

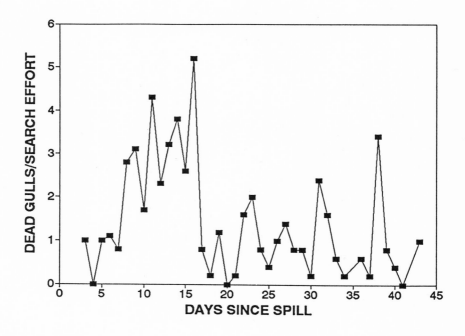

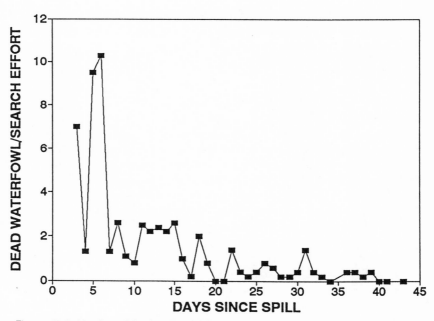

Figure 13.2. Number of (top) gulls and (bottom) waterfowl found dead during search-and-rescue operations in the Arthur Kill and environs (from Louis Berger & Associates, Inc., 1991).

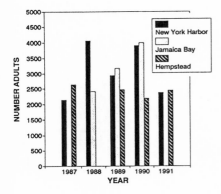
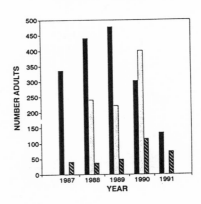
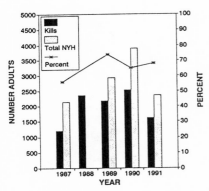
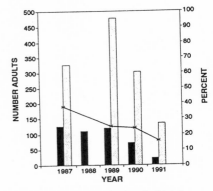

Figure 13.3. Abundance of nesting herring (left) and great black-backed gulls (right) in the metropolitan waterways and wetlands of New York City, 1987–91 (from Downer and Liebelt 1990; Liebelt pers. comm.).

Of the herring and great black-backed gull populations utilizing New York Harbor, a significant proportion nested on islands in the kills (Fig. 13.3). Loss of abundance of great black-backed gulls in the kills reflected harborwide losses since 1989.

Abundance of gulls nesting on Pralls Island in the Arthur Kill shrank somewhat in 1990, especially along the eastern salt marsh shore of the island ($p < .001$). The western shore of primarily cobble and sand beaches that face the Kill continued to support approximately four hundred pairs of gulls since the mid 1980s (Fig. 13.4). Measures of avian productivity, including egg size, clutch size, and hatching success (Table 13.2) showed that herring gulls were relatively insensitive to the adverse effects of habitat degradation caused by the oil spills of 1990.

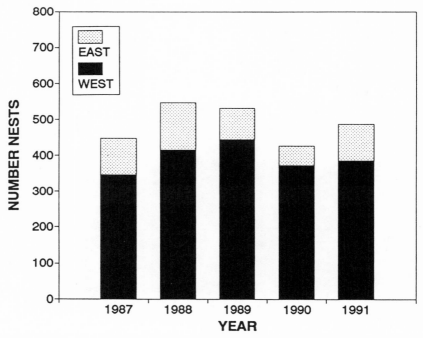

Figure 13.4. Abundance and spatial distribution of herring and great black-backed gulls nesting on the east and west shores of Pralls Island, 1987–91.

Long-legged Waders

Since the late 1970s, herons, egrets, and ibises (waders) have at an accelerating rate claimed islands in New York Harbor for the rare nesting habitat they provide. The proportion of New York coastal waders utilizing the kills grew from 3 percent in 1977 to 30 percent by 1989 (Burger, Parsons, and Gochfeld in press). Furthermore, at its 1991 level of thirteen hundred pairs, the kills wader population represented 20 to 30 percent of all the waders that bred along the coasts of New Jersey, New York, and Connecticut (Downer and Liebelt 1990; NJDEPE 1990; M. Bull [Conn. Audubon Society] pers. comm.). Because of their abundance and diversity, and because of their critical importance to regional avian biodiversity, the waders of New York Harbor represent a significant wildlife community in this urban estuary.

The populations nesting in the kills have been the focus of a multidimensional research project conducted by the Manomet Bird Observatory since 1985. While unfortunate, the close proximity of the 1990 Exxon spill to heron feeding and breeding sites and the existence of detailed ecologi-

Table 13.2. Herring Gull Egg Production and Hatching Success of Three-Egg Clutches on Pralls Island, 1989–91

	1989	1990	1991
Egg length (mm)	70.7 ± 2.4 (35)[1]	70.8 ± 2.7 (30)[1]	71.1 ± 1.9 (11)[1]
Egg width (mm)	49.7 ± 1.9 (35)[1]	50.0 ± 1.9 (30)[1]	49.8 ± 0.8 (11)[1]
Clutch size	2.7 ± 0.6 (256)[1a]	2.4 ± 0.7 (25)[1]	2.6 ± 0.7 (27)[1]
No. hatched	—	2.3 ± 1.1 (21)[1]	1.8 ± 1.2 (17)[1]

NOTE: Given are mean ± sd (N nests). Like superscripts are not different.
[a]Data from 1988 (1989 data unavailable).

Table 13.3. Abundance of Active Wader Nests on Shooters Island (SI), Pralls Island (PI), and the Isle of Meadows (IM) during late-May surveys 1986–91

Year	SI	PI	IM	Total
1986	244	627	0	871
1987	205	419	+	624
1988	234	167	832	1233
1989	208	212	517	937
1990	293	394	469	1156
1991	704	273	342	1319

SOURCE: Parsons 1990, 1991.
NOTE: + = birds were present but not counted.

cal data on these populations provided a unique opportunity to examine parameters of avian ecology both prior to and following the spill (Parsons 1990, 1991).

Abundance

Most wader populations in the kills continued to increase in size in 1990 and 1991, illustrating the rich ecological potential of the estuary (Table 13.3). Despite strong overall wader abundance, however, glossy ibis (*Plegadis falcinellus*), a sensitive tidal species, experienced an unprecedented drop (42%) in abundance in 1991 (Fig. 13.5). Black-crowned night herons (*Nycticorax nycticorax*), an opportunistic species, grew in abundance by 34 percent.

In addition, disruption of ecological organization at the community level was evident as two of three colony sites in the kills were largely abandoned in favor of Shooters Island, which formerly supported only 20 to 30 percent of the kills' waders. In 1991, Shooters Island provided nesting habitat for more than 50 percent of the area's waders (Table 13.3).

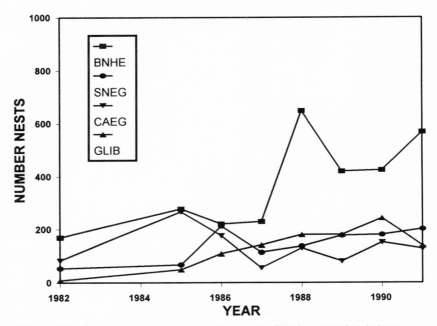

Figure 13.5. Abundance of dominant wader species (black-crowned night heron [BNHE], snowy egret [SNEG], cattle egret [CAEG], glossy ibis [GLIB]) nesting in the kills, 1982–91 (from Parsons 1991).

Significant relocation of birds to Shooters Island indicated a loss of colony site fidelity to the other islands in the kills. In 1990, tidal waders fledged many fewer nestlings from Pralls Island and Isle of Meadows than in previous years. The near-complete absence of small egrets and ibises from Isle of Meadows and reduced populations on Pralls Island in 1991 may have reflected their poor reproductive performance at these sites in 1990. The concentration of breeding effort at any single colony site is less desirable than a more uniform representation at all three islands because the birds are more vulnerable to site-specific and density-dependent problems (such as parasitic infestations, infectious diseases, vandalism, and nest-site competition).

Nesting Ecology

In 1990, nest initiation was significantly late at all three colony sites. Fewer than one-third of expected nesting attempts occurred in April (Parsons 1990). However, subsequent mean hatch dates and nesting

Table 13.4. Mean Hatch Dates ($\pm$ sd (da)) for Waders Nesting in the Arthur Kill, 1986–91

Species	Pre-1990	1990	1991
Black-crowned night-heron	26 May $\pm$ 16.8 (103)[1]	24 May $\pm$ 15.3 (59)[1]	2 Jun $\pm$ 5.2 (56)[2]
Snowy egret	30 May $\pm$ 13.7 (127)[1]	2 Jun $\pm$ 17.9 (48)[1]	2 Jun $\pm$ 12.1 (26)[1]
Cattle egret	4 Jun $\pm$ 14.3 (70)[1]	15 Jun $\pm$ 20.0 (20)[2]	30 May $\pm$ 5.1 (9)[1]
Glossy ibis	23 May $\pm$ 9.2 (91)[1]	21 May $\pm$ 4.3 (60)[1]	6 Jun $\pm$ 5.3 (14)[2]

SOURCE: Parsons, 1990, 1991.
NOTES: Hatch dates of estimated renesters are excluded.
For each species, unlike superscripts indicate $p < .05$.

synchrony did not differ except for cattle egrets (*Bubulcus ibis*), an upland foraging species (Table 13.4).

In 1991, mean hatch dates of both black-crowned night herons and glossy ibises were later than expected (Table 13.4). However, cattle egrets returned to a prespill timetable. The nesting chronology of snowy egrets (*Egretta thula*), a salt marsh–dependent species, did not differ over the entire 6-year-study period (Table 13.4).

Although generalist and upland foraging species (i.e., black-crowned night heron, cattle egret) were relatively unaffected by loss of tidal habitat in 1990, species foraging primarily in salt marshes and tidal mud flats (i.e., snowy egret, glossy ibis) showed lowered reproductive success in several parameters. Throughout the study period, snowy egrets consistently hatched 75 to 85 percent of their four-egg clutches (Table 13.5). Yet since 1989, the ability of snowy egrets to fledge hatchlings fell dramatically. During the prespill seasons of 1986 to 1989, snowies produced approximately 1.5 young per nest. By 1991, production fell to 0.3 young per nest. Similarly, glossy ibis production fell from 1.3 young per nest to 0.4 young per nest in 1991 (Table 13.5). In contrast, black-crowned night heron and cattle egret chick production did not vary over the study period (Table 13.5).

Only tidal waders experienced increased nestling mortality in 1990 compared with prespill data (Table 13.6). Snowy egret losses were the heaviest. In 1991, tidal waders continued to fail, with almost 90 percent of all snowy egret hatchlings perishing before 3 weeks of age. In addition, more black-crowned night heron chicks failed to survive to 3 weeks in 1991 compared with previous data (Table 13.6). Most tidal wader losses (snowy egret, glossy ibis) in both postspill years were attributed to starvation as reflected by lower growth rates for some species (Table 13.7). However, most night heron and cattle egret losses postspill were attributed to predation (Parsons 1990, 1991).

Table 13.5. Productivity of Herons, Egrets, and Ibises Nesting in the Arthur Kill, 1986–91

Species	Prespill (1986–89)	Postspill 1990	Postspill 1991	p
Black-crowned night heron				
Clutch	3.0 ± 0.7	3.3 ± 0.8	3.1 ± 0.9	ns
Hatched/clutch	0.69 ± 0.38	0.72 ± 0.39	0.62 ± 0.43	ns
Fledged/hatch	0.64 ± 0.43	0.57 ± 0.37	0.43 ± 0.39	.0347
Young fledged	1.7 ± 1.3	1.6 ± 1.2	1.2 ± 1.2	ns
Snowy egret				
Clutch	3.9 ± 0.7	4.0 ± 0.7	4.1 ± 0.7	ns
Hatched/clutch	0.74 ± 0.32	0.86 ± 0.21	0.84 ± 0.21	ns
Fledged/hatch	0.47 ± 0.43	0.22 ± 0.28	0.09 ± 0.21	.0320
Young fledged	1.5 ± 1.3	0.7 ± 0.9	0.3 ± 0.8	.0387
Cattle egret				
Clutch	3.3 ± 0.8	3.3 ± 0.8	3.7 ± 0.9	ns
Hatched/clutch	0.80 ± 0.30	0.82 ± 0.18	0.57 ± 0.42	ns
Fledged/hatch	0.51 ± 0.42	0.47 ± 0.43	0.32 ± 0.36	ns
Young fledged	1.4 ± 1.2	1.2 ± 1.1	0.8 ± 0.8	ns
Glossy ibis				
Clutch	3.1 ± 0.6	3.1 ± 0.6	3.1 ± 0.7	ns
Hatched/clutch	0.72 ± 0.38	0.67 ± 0.39	0.46 ± 0.49	ns
Fledged/hatch	0.48 ± 0.43	0.32 ± 0.36	0.13 ± 0.26	.0612
Young fledged	1.3 ± 1.2	0.9 ± 1.0	0.4 ± 0.9	ns

Source: Parsons 1990, 1991.
Notes: Given are x ± sd per nest.
Comparisons between 1990 and 1991 data were made with Wilcoxon 2-sample test. ns = not significant.

Foraging Ecology

An examination of foraging flights from three colony sites in the kills over the period 1986 to 1990 showed that tidal waders undertook fewer flights in 1990 compared with prespill data (Parsons 1990). The decreased foraging flight rates but unchanging flight directions were best explained by tidal waders spending greater than expected time periods at foraging sites (Parsons 1990). Although cattle egrets showed no variation in flight rate, minor shifts in flight direction were documented.

The scenario of tidal waders selecting traditional foraging sites despite the loss of prey populations (Burger, Brzorad, and Gochfeld 1991; Parsons, unpub. data) as a result of oil contamination was further supported by wetland survey data. No differences in wader abundance (pre- and postspill) at several tidal wetlands were observed (Parsons 1990). Furthermore, diet analysis showed that snowy egrets, while not eschewing a saltwater diet, brought more estuarine shrimp to nestlings in 1990 than in previous years. In contrast, black-crowns fed more heavily on human refuse and less on estuarine fish in 1990 (Fig. 13.6).

Summary

Despite a long history of environmental degradation and wetlands alteration, the Arthur Kill continued to provide important breeding and feeding habitat for many waders through 1991. Few long-term field studies exist; however, one study of seabirds suggests that bird populations may be adversely affected by oil contamination several years following a spill (Fry et al. 1986).

Our study showed that species feeding in tidal habitats in the kills experienced lowered reproductive success and disrupted foraging ecologies in the breeding season immediately following the 1990 Exxon spill. Most losses were attributed to starvation as a result of tidal-habitat degradation and diminished prey populations. Analysis of foraging data showed that some species were relatively inflexible in adjusting to pollution events (e.g., by selecting new feeding sites). Nontidal or generalist species were relatively unaffected (Parsons 1990).

Two years after an unprecedented volume of oil contaminated their habitats, tidal waders continued to show signs of ecological stress. In 1991, we found changes in colony site use, species abundance, and reproductive output that suggest destabilization may have occurred at the community level (Parsons 1991). Of special concern is the plight of the snowy egret, a wader inextricably tied to estuarine marshes. Yet despite evidence that food resources are inadequate to fledge nestlings, all waders continued to lay and hatch expected numbers of eggs, indicating that the potential exists to recover losses.

Waterfowl

New York Harbor provides habitat to thousands of wintering ducks and geese (Fig. 13.7). However, annual aerial surveys conducted in January never show more than one hundred to three hundred waterfowl utilizing the Arthur Kill (NJDEPE 1991).

Waterfowl in the waterways extending from the mouth of the Raritan River through the kills and Newark Bay, and to the Hackensack and Passaic rivers constitute only 3 to 30 percent of the total northern New Jersey waterfowl populations (Fig. 13.8). In contrast, waterfowl counted in the metropolitan waterways of New York City constitute 25 to 44 percent of waterfowl surveyed in the Hudson River valley and Long Island (NYSDEC 1991).

Table 13.6. Proportion of Wader Nestlings Failing to Survive to 20 Days per Hatched Chicks in Study Nests on Isle of Meadows and Pralls Island, 1986–91

Species	Pre-1990			1990			1991		
	dead	hatch	%	dead	hatch	%	dead	hatch	%
Black-crowned night-heron	33[1]	205	16%	33[1]	150	22%	46[2]	115	40%
Snowy egret	101[1]	243	42	130[2]	180	72	77[3]	88	88
Cattle egret	74[1]	167	44	24[1]	53	45	11[1]	18	61
Glossy ibis	73[1]	197	37	63[2]	132	48	17[3]	23	74

SOURCE: Parsons 1990, 1991.

NOTE: For each species, unlike superscripts indicate differences are significant (p < .05).

Table 13.7. Growth Indices of Wader Nestlings in the Arthur Kill, 1989–91

	1989	1990	1991
Black-crowned night heron			
A	10.3 ± 0.3 (3)[1]	11.8 ± 1.2 (11)[2]	10.8 ± 1.0 (5)[1]
C	6.3 ± 2.1 (3)[1]	11.2 ± 1.9 (10)[2]	8.6 ± 1.1 (4)[1]
Snowy egret			
A	6.1 ± 0.5 (5)[1]	6.6 ± 0.9 (8)[1]	5.5 ± 0.5 (5)[2]
C	6.7 ± 0.6 (5)[1]	5.5 ± 1.9 (5)[2]	4.1 ± 0.9 (3)[3]
Cattle egret			
A	5.4 ± 1.4 (5)[1]	5.6 ± 0.2 (5)[1]	5.2 ± 0.4 (2)
C	6.8 ± 0.7 (3)[1]	6.7 ± 0.5 (3)[1]	4.9 (1)
Glossy ibis			
A	7.7 ± 1.3 (4)[1]	7.8 ± 1.0 (9)[1]	7.0 ± 0.1 (2)
C	7.0 ± 0.7 (4)[1]	6.2 ± 1.5 (7)[1]	7.5 (1)

SOURCE: Parsons 1991.
NOTES: Given are mean ± sd (N nestlings) rates (mass = [m] tarsus + b) for the first-hatched (A) chick and third-hatched (C) chick.
Unlike superscripts indicate $p < .05$.

Almost half of the 679 dead birds recovered during the 6 weeks following the Exxon oil spill were waterfowl (Table 13.1). (Four other birds were not included in the Table 13.1 because they were not waterbirds.) Since 1989, mallards have never numbered more than 100 individuals during midwinter surveys (NJDEPE 1991), yet 135 mallards were among the 283 ducks and geese killed following the Exxon spill (Table 13.1). Waterfowl recoveries peaked during the first week following the spill, whereas gull recoveries were distributed somewhat later (13.2). This difference may reflect that gulls showed greater site fidelity than the waterfowl or that waterfowl spent more time on the water surface than gulls, or that they were more vulnerable to immediate oiling.

January abundance data from the eastern part of New York Harbor show that, with the possible exception of mallards, dominant waterfowl species were unaffected by the Exxon oil spill (Fig. 13.7). Similarly, wintering waterfowl populations in the western part of the estuary appeared unaffected, except possibly scaup (*Aythya spp.*) (Fig. 13.8), which feed primarily on mussel spat.

Although wintering waterfowl populations in New York Harbor showed minimal response to massive oil contamination of the kills, subsequent nesting populations of ducks in the Arthur Kill were much lower than expected ($p < .05$). In 1991, duck populations on Pralls Island were still low (Table 13.8). In contrast, Canada geese have continued to build nesting populations in the kills. Despite losing 16 individuals in the

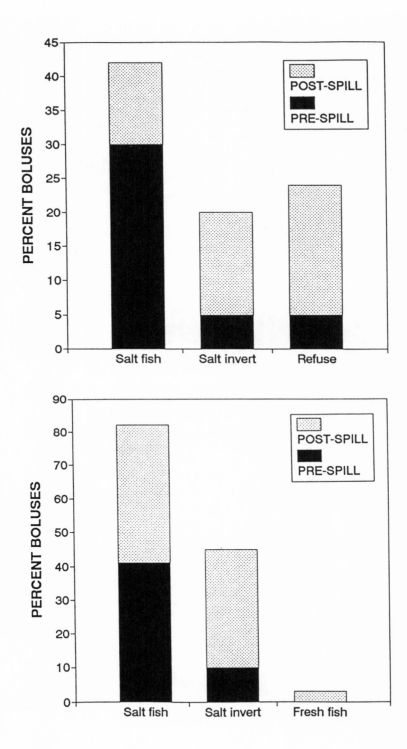

aftermath of the Exxon spill, the breeding goose population on Pralls Island more than quadrupled in 1990 and grew again by 78 percent in 1991 (Table 13.8). Furthermore, clutch size of Canada goose was not different from expected in 1990 compared with previous years' data (Table 13.9).

Conclusions

Members of the rich community of aquatic birds utilizing the Arthur Kill and Kill van Kull showed varied responses to the massive oil contamination that degraded tidal habitats in 1990. Some species, such as the snowy egret and glossy ibis, experienced precipitous declines in the number of young successfully raised to fledging. Although abundance of these tidal species remained relatively stable, their continued inability to fledge nestlings 2 years after the spill threatens their persistence in the kills.

In contrast, opportunistic species such as the black-crowned night heron and cattle egret showed minimal adverse effects. Canada goose have also succeeded in building a strong breeding population in the kills since 1989. Species showing intermediate responses, such as the double-crested cormorant, and gull and duck species, apparently were adversely affected in the short term but showed adjustment, or declines were attributed to factors other than the oil spills of 1990. In general, diet specificity and degree of dependence on oiled habitats determined the magnitude of response from the kills' aquatic birds.

Recommendations

Manomet Bird Observatory began a long-term study of the kills' avifauna 4 years prior to the harbor's largest oil spill on record. Because of this established database, we were able to interpret pre- and postspill measures of avian abundance, productivity, and habitat use with relative confidence. In addition, our study was broad enough to allow comparisons between bird populations dependent on tidal resources and populations that make use of other habitats. We therefore were able to approach the problem of assessing impacts from both temporal (before–after) and spatial (tidal–nontidal) standpoints. These are extremely powerful attributes

Figure 13.6. Major diet items of (top) black-crowned night herons and (bottom) snowy egrets during prespill (1988–89) and postspill (1990) seasons. Given are percentage occurrences in examined boluses (from Parsons 1990).

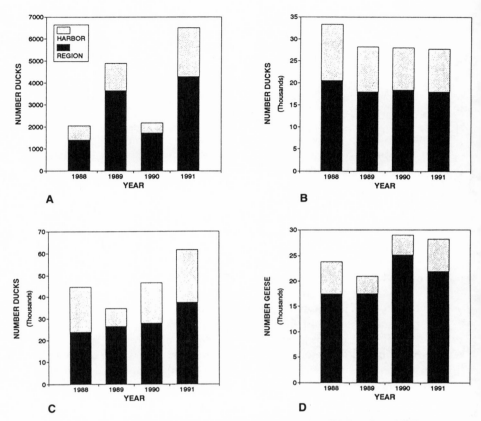

Figure 13.7. Abundance of wintering waterfowl, (A) mallard, (B) American black duck, (C) scaup (*Aythya* spp.), (D) Canada goose, in the eastern part of New York Harbor, 1988–91. Given are numbers of individuals counted during January aerial surveys (from NYSDEC 1991). Adult counts include immediate area of spill site ("Harbor") and other water bodies in the general region ("Region").

in a field study and are without precedent in other examinations of birds and oil spills.

Investment in biological monitoring is advantageous when the wildlife populations in the area are valued and the likelihood of environmental contamination is relatively high. In addition, biological monitoring of higher organisms such as birds is a useful and cost-effective tool for habitat management. Species whose ecology shows a close and sensitive relationship to their foraging habitats, such as the snowy egret have not only the ability to reveal direct threats from contaminated food chains but can also be sensitive indicators of habitat loss. Coastal areas and

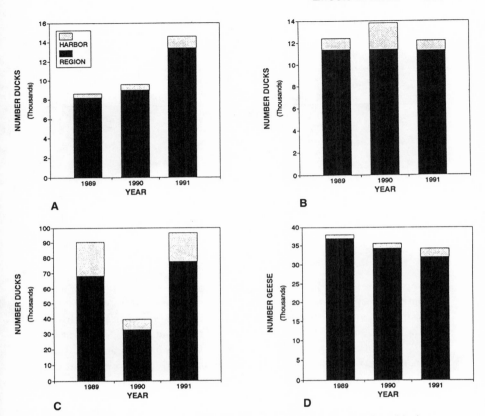

Figure 13.8 Abundance of wintering waterfowl, (A) mallard, (B) American black duck, (C) scaup (*Aythya* spp.), (D) Canada goose, in the western part of New York Harbor, 1989–91. Given are numbers of individuals counted during January aerial surveys (from NJDEPE 1991). Adult counts include immediate area of spill site ("Harbor") and other water bodies in the general region ("Region").

Table 13.8. Abundance of Nesting Waterfowl on Pralls Island, 1987–91

Year	Canada goose	American black duck	Mallard	Gadwall	Total ducks
1987	1	6	2	0	8
1988	3	10	6	13	29
1989	2	1	2	13	16
1990	9	1	1	1	3
1991	16	0	4	1	5

NOTE: Given are total active nests.

Table 13.9. Waterfowl Productivity on Pralls Island, 1988–90

| | Clutch size | | |
	1988	1989	1990
Canada goose	5.5 ± 0.7 (2)[1]	4.5 ± 0.7 (2)[1]	6.0 ± 1.1 (9)[1]
Total ducks	8.4 ± 2.0 (20)[1]	8.4 ± 3.9 (9)[1]	

NOTES: Given one mean ± sd (N nests) clutch size for Canada goose and for mallard, American black duck, and gadwall combined.
Like superscripts are not different.

other aquatic ecosystems vulnerable to chemical pollution should implement biological monitoring plans as proactive strategies for limiting habitat degradation and protecting wildlife populations.

Acknowledgments

This report summarizes 6 years of research conducted in conjunction with the Harbor Herons Project, a collaboration of private and public agencies whose principal goal is to conserve newly established heron populations and their habitats in New York Harbor. Manomet Bird Observatory (MBO), in partnership with the New York City Audubon Society, New York City Department of Parks and Recreation (Natural Resources Group), Trust for Public Land, and New York State Department of Environmental Conservation, has conducted scientific studies of the wader populations in New York City since 1985 to provide ecological data critical to the wildlife conservation, habitat protection, and public education goals of the project.

The initiation and land protection accomplishments of the Harbor Herons Project are due in large part to the efforts of the founding organization, the New York City Audubon Society, and especially to the vision and leadership of former director Albert Appleton and to Scottie Jenkins.

MBO's studies have been accomplished with the help of many dedicated individuals. I thank research assistants Amanda McColpin, Kathleen Vance, and Alan Maccarone. Field assistance was provided by Beverly DeAngelis, Diana Yates, Michael Morse, Marina Arandjelovic, and John Brzorad. In addition, numerous volunteers have contributed to the project.

Members of the Harbor Herons Project Scientific Advisory Board have provided important scientific guidance. I am grateful to Drs. Donald McCrimmon, Joanna Burger, and R. Michael Erwin. I thank Dr. Ward Stone for toxicological analyses and for his interest in the project. I am especially grateful to Dennis Heinemann for analytical and statistical assistance.

This work was supported in part through a contract with the New Jersey Department of Environmental Protection under the New Jersey Spill Compensation Fund (SCF 685), with monies supplied to the fund by Exxon Company, U.S.A. Additional contract oversight was provided by the New York State Department of Environmental Conservation, the New York City Departments of Environmental Protection and Parks and Recreation, the Elizabeth (N.J.) Department of Health, Welfare and Housing, the National Oceanic and Atmospheric Administration, and the U.S. Fish and Wildlife Service. Work undertaken in 1991 was partially supported by the Arthur Kill Trust Fund, a Court Registry Account established pursuant to the disposition of *United States v. Exxon Corporation*.

We are especially grateful to the following organizations for their farsighted and long-term support of the Harbor Herons Project: New York City Audubon Society, Geraldine R. Dodge Foundation, Eppley Foundation, New York State Department of Environmental Conservation Return-a-Gift Program, U.S. Environmental Protection Agency, and MBO members.

Literature Cited

Berger, L., & Associates. 1991. The Arthur Kill oil discharge study. Final report submitted to New Jersey Department of Environmental Protection and Energy.

Boersma, P. D. 1988. Weathered crude oil effects on chicks of fork-tailed storm petrels (*Oceanodroma furcata*). *Arch. Environ. Contam. Toxicol.* 17:527–531.

Buckley, P. A., and F. G. Buckley. 1984. Expanding double-crested cormorant and laughing gull populations on Long Island, NY. *Kingbird* 34:146–155.

Burger, J., J. Brzorad, and M. Gochfeld. 1991. Immediate effects of an oil spill on behavior of fiddler crabs (*Uca pugnax*). *Arch. Environ. Contam. Toxicol.* 20:404–409.

Burger, J., K. C. Parsons, and M. Gochfeld. In press. Avian populations and environmental degradation in an urban river: The kills of New York and New Jersey. In J. A. Jackson, ed., *Avian Conservation* Madison: University of Wisconsin Press.

Butler, R. G. 1988. Impact of sublethal oil and emulsion exposure on the reproductive success of Leach's storm petrels: Short- and long-term effects. *J. Appl. Ecol.* 25:125–143.

Clark, R. B. 1984. Impact of oil on seabirds. *Environ. Pollut. Ser. A Ecol. Biol.* 33:1–22.

Crocker, A. D., J. Cronshaw, W. N. Holmes. 1974. The effect of crude oil on intestinal absorption in ducklings (*Anas platyrhynchos*). *Environ. Pollut.* 7:165–177.

Downer, R. H., and C. E. Liebelt. 1990. *1989 Long Island Colonial Waterbird and Piping Plover Survey*. New York State Department of Environmental Conservation report.

Farrington, J. W. 1977. Oil pollution in the coastal environment. In *Estuarine pollution control and assessment*. Report submitted to the U.S. Environmental Protection Agency.

Freedman, B. 1989. *Environmental Ecology: The Impact of Pollution and Other Stresses on Ecosystem Structure and Function*. San Diego, Calif.: Academic Press.

Fry, D. M., J. Swenson, L. A. Addiego, C. R. Grau, and A. Kang. 1986. Reduced reproduction of wedge-tailed shearwaters exposed to weathered Santa Barbara crude oil. *Arch. Environ. Contam. Toxicol.* 15:453–463.

Hackl, E., and J. Burger. 1988. Factors affecting piracy in herring gulls at a New Jersey landfill. *Wilson Bull.* 100:424–430.

Hartung, R. 1967. Energy metabolism in oil-covered ducks. *J. Wildl. Manage.* 31:798–804.

Hartung, R., and G. S. Hunt. 1966. Toxicity of some oils to waterfowl. *J. Wildl. Manage.* 30:564–570.

Hatch, J. J. 1984. Rapid increase of double-crested cormorants nesting in southern New England. *Am. Birds* 38:984–988.

Hoffman, D. J., and P. H. Albers. 1984. Evaluation of potential embryotoxicity and teratogenicity of 42 herbicides, insecticides, and petroleum contaminants to mallard eggs. *Arch. Environ. Contam. Toxicol.* 13:15–17.

International Council for Bird Preservation. 1991. *ICBP/RSPB Gulf Oil Spill Team Reports*. Cambridge: ICBP.

Kadlec, J. A., and W. H. Drury. 1968. Structure of the New England herring gull population. *Ecology* 49:644–676.

King, K. A., and C. A. Lafever. 1979. Effects of oil transferred from incubating gulls to their eggs. *Mar. Poll. Bull.* 10:319–321.

Miller, D. S., D. B. Peakall, and W. B. Kinter. 1978. Ingestion of crude oil: Sublethal effects in herring gull chicks. *Science* 199:315–317.

Monaghan, P. 1979. Aspects of the breeding biology of herring gulls *Larus argentatus* in urban colonies. *Ibis* 121:475–481.

New Jersey Department of Environmental Protection and Energy (NJDEPE). 1990. Colonial waterbird surveys. NJDEPE Division of Nongame report.

———. 1991. Aerial waterfowl surveys. NJDEPE Division of Fish and Game report.

New York State Department of Environmental Conservation (NYSDEC). 1991. Aerial waterfowl surveys. NYSDEC Division of Fish and Game report.

Ohlendorf, H. M., E. E. Klaas, and T. E. Kaiser. 1978. Organochlorine residues and eggshell thinning in wood storks and anhingas. *Proc. Colonial Waterbird Group* 1:185–195.

Parsons, K. C. 1990. Aquatic birds of the Arthur Kill: Short-term impacts following major oil spills. Final report submitted to New Jersey Department of Environmental Protection and Energy.

———. 1991. Aquatic birds of the Arthur Kill: Second-year impacts to Ciconiiformes following major oil spills. Draft interim report submitted to New Jersey Department of Environmental Protection and Energy.

Parsons, K. C., A. D. Maccarone, and J. Brzorad. 1991. First breeding record of double-crested cormorant (*Phalacrocorax auritus*) in New Jersey. *Records of NJ Birds* 17:51–52.

Peakall, D. B., D. S. Miller, and W. B. Kinter. 1983. Toxicity of crude oils and their fractions to nestling herring gulls. Physiological and biochemical effects. *Mar. Envrion. Res.* 8:63–71.

Piatt, J. F., C. J. Lensink, W. Butler, M. Kendziorek, and D. R. Nysewander. 1990. Immediate impact of the *"Exxon Valdez"* oil spill on marine birds. *Auk* 107:387–397.

Stickel, L. F., and M. P. Dieter. 1979. Ecological and physiological/toxicological effects of petroleum on aquatic birds. U. S. Fish and Wildlife Services Program, FWS 1035–79/23.

Szaro, R. C. 1977. Effects of petroleum on birds. *Trans. N. Am. Wildl. Nat. Resour. Conf.* 42:374–381.

14. Biological Effects: Marine Mammals and Sea Turtles

Romona Haebler

In the description of the Arthur Kill biological features presented in Chapter 1, marine mammals and sea turtles are not discussed since they are not regular residents of this area. However, marine turtles, seals and cetaceans are occasionally sighted in the Arthur Kill, and they are common inhabitants of the New York bight as well as most U.S. coastal waters. Although the spill in the Arthur Kill did not harm these species, other spills have had serious effects. For example, only 9 months before the Arthur Kill spill occurred, the *Exxon Valdez* ran aground Bligh Reef in Prince William Sound, Alaska, prime habitat for marine mammals, and spilled over 11 million gallons of crude oil. Oil spills can occur at any time in waters where oil is either transported or drilled. In 1979, a massive blowout of the Ixtoc I oil well in the Gulf of Mexico, critical habitat to sea turtles and home range for many marine mammals, released over a half million gallons of oil *per day* for months (Hall, Belisle, and Sileo 1983).

All spills are different, varying in type and amount of oil spilled, species exposed, and geographic and atmospheric conditions. It is important to understand as much as possible about both the natural history and characteristics of various species and the specific effects oil has on wildlife. Doing so improves the ability to extrapolate from one spill to another and improves prediction of types and severity of effects to wildlife. This chapter presents an overview of the biological effects of oil on marine mammals and sea turtles. For a more detailed review of the biological effects of oil on marine mammals and sea turtles, see Geraci and St. Aubin 1990 and Hutchinson and Simmonds 1991.

Animals at Risk

Though it is commonly accepted that oil spilled into the marine environment may cause adverse biological effects on marine mammals and turtles, the level of risk varies with the specific anatomical and physiological characteristics and habitat requirements of a given species. Marine mammals are a subset of Mammalia, which have evolved unique characteristics that allow survival within an aquatic system. The five major groups of marine mammals include pinnipeds (seals, sea lions, and walruses), cetaceans (whales and dolphins), mustelids (sea otters), sirenians (manatees and dugongs), and ursids (polar bears). Since these five groups did not evolve from a common phylogenetic line, they vary tremendously in both biological and physical characteristics. All of these marine-mammal species, with the exception of dugongs, inhabit U.S. coastal waters. Five species of sea turtle occur in these same waters: Kemp's ridley (*Lepidochelys kempi*), loggerhead (*Caretta caretta*), green turtle (*Chelonia mydas*), hawksbill (*Eretmochelys imbricata*), and leatherback (*Dermochelys coriacea*). Since these five species of reptiles have similar biological and physical characteristics, their relative risk of biological effects due to oil exposure is similar.

Though they do not live in the oceans, many terrestrial mammals may also suffer biological effects following exposure to oil spilled into the marine environment. These include species that take their food from the sea (e.g., river otter and mink) and species that sometimes feed on carrion (e.g., coyote and bear).

Toxicity of Oil to Mammals and Turtles

Physical and chemical properties of crude or refined petroleum products determine their toxicity and persistence in the marine environment. Acute toxicity correlates directly with concentration of light aromatic hydrocarbons (Neff and Anderson 1981; National Academy of Sciences 1983). These low-molecular-weight compounds volatilize rapidly at the air–sea interface, where all marine mammals and sea turtles must come to breathe. Exposure to noxious compounds may cause irritation to eyes or mucous membranes. Though chemical components of heavier petroleum products are not as acutely toxic, the physical properties of these more viscous oils which make them thick and sticky pose risks to wildlife through contamination of pelage, clogged breathing passages, and fouling of baleen. Also, heavier oils persist in the environment and may contaminate beaches and other haul-out sites (places where animals leave the water for the shore). The severity of risk associated with physical contact depends on texture of the exposed body surface, frequency and duration of exposure, and characteristics of the oil (Engelhardt 1978).

The chemical and physical properties of oils change with time due to the process of weathering. As a result, chemical toxicity usually decreases due to evaporation, dissolution, and so on, but the physical properties of weathered oil may become even more harmful to wildlife. As oil is emulsified, "mousse" or water-in-oil emulsions that can be extremely thick and tenacious are formed. This material creates a particularly hazardous problem for animals, by coating them and by contaminating beaches or rookeries.

Mammals and turtles are exposed to oil by three main routes: physical contact (coating, inhalation, and ingestion (both direct ingestion of oil or tar balls and ingestion of contaminated food). Most of this exposure occurs when animals contact oil at the sea surface or in areas where oil has fouled the shoreline. Through processes of dissolution and dispersion, oil enters the water column, but exposure from this source is not thought to be a serious threat to wildlife since concentrations are usually relatively low (Engelhardt 1987). Oil may enter sediment, where it can be taken up by a variety of benthic animals and therefore pose an exposure risk to mammals and turtles that feed on these organisms. The behavior of oil spilled into an ice-covered marine environment has some unique characteristics and can be a serious threat to marine mammals. For information about this see Clark and Finley 1982 and Geraci 1990.

Biological Effects in Marine Mammals

Pinnipeds

The thirty-four species of pinnipeds can be divided into three major categories: hair seals (family Phocidae), fur seals and sea lions (family Otaridae), and walrus (family Odobenidae). Two characteristics that vary among these groups can affect their relative risk to adverse oil effects: pelage (e.g., fur seals by physical coating) and feeding habits (e.g., walrus by ingestion of contaminated benthic animals). Otherwise, risks from exposure to oil are fairly consistent across species, varying primarily with geographic distribution. Pinnipeds inhabit U.S. coastal waters along the entire Pacific coast and the Atlantic coast as far south as New Jersey. No pinnipeds occur in the Gulf of Mexico.

Physical Contact Pinnipeds have exceptionally large, protruding eyes, which may make them particularly susceptible to toxic volatile-aromatic compounds. Irritation of eyes and mucous membranes is reported following both experimental exposure to oil slicks (Smith and Geraci 1975; Geraci and Smith 1976) and oil spills in the natural environment (Lillie 1954). Since volatile compounds evaporate rapidly (usually within 24 to 48 hours), this risk should decrease with time following a single release. However, hazards associated with this mechanism of toxicity will be more severe during a well blowout or other event in which oil is discharged over an extended time.

The thick pelage of the northern fur seal provides insulation necessary for the animal to maintain thermoregulation. Contamination with petroleum hydrocarbons removes natural oils and increases heat transfer, thus causing a higher metabolic demand. Pinnipeds other than fur seals depend primarily on their thick blubber layer for insulation and therefore are at much lower risk from thermal effects associated with fouling of the pelage.

An exception, however, are newborn phocid seals, which depend on lanugo, a thick, prenatal fur coat, for insulation. Unlike most marine mammals, pinnipeds give birth on land, where newborns then spend several days while the mothers forage. Rookeries are commoly located in rocky outcroppings, which may readily entrap weathered oil, thus increasing risk of exposure.

Though pinnipeds coated in oil can usually rid themselves of contamination within a short time (ranging from days to weeks) (Le Boeuf 1971; Geraci and Smith 1976), it has been suggested that fouling with thick, heavy oil may impair swimming ability, even leading to drowning (Warner 1969; Davis and Anderson 1976).

Inhalation Experimental exposure of six ringed seals (*Phoca hispida*) to oil-covered water resulted in blood and tissue levels of petroleum hydrocarbons in the range of several parts per million (Engelhardt, Geraci, and Smith 1977). Since no oil was identified in the digestive tract on postmortem examination, the investigators concluded that the petroleum residues entered the body via inhalation across respiratory epithelium. Two of the six seals showed histological evidence of renal damage, and one had fatty degeneration of the liver (Smith and Geraci 1975). Concentrations of hydrocarbons in these tissues correlated with occurrence of lesions.

A common pathological finding in wild populations of pinnipeds is parasitism with lungworms and associated secondary-disease processes such as pneumonia. Exposure to an inhalation toxicant or irritant may be especially dangerous to individual animals that have preexisting pulmonary disease.

Ingestion Experimental studies on ringed seals using oral doses of radiolabeled petroleum hydrocarbons (Engelhardt, Geraci, and Smith 1977; Engelhardt 1978, 1982) show that these compounds are absorbed into the bloodstream and distributed to various body tissues; concentrations are highest in liver and blubber. Both renal and hepatic/biliary enzyme systems appear to be effective in metabolizing petroleum hydrocarbons to polar metabolites for excretion (Engelhardt 1978, 1982; Addison and Brodie 1984; Addison et al. 1986), primarily through induction of mixed-function oxidases (MFO). The enzyme aryl hydrocarbon hydroxylase, identified in both liver and kidney tissue of ringed seals, could be induced in vitro by exposure to crude oil (Engelhardt 1982). In grey seals (*Halichoerus grypus*) and harbor seals (*Phoca vitulina*) (Addison et al. 1986) the hepatic MFO system is more developed in adults than in the young (Addison and Brodie 1984), suggesting that due to metabolic incompetence petroleum exposure may be more harmful to newborn animals than to adults.

Residues of petroleum hydrocarbons are in lipid-rich tissues of several species of pinnipeds (Risebrough et al. 1978; Geraci and St. Aubin 1985), indicating that these species have been exposed to oil and that these compounds are not completely eliminated from the body and therefore may pose a long-term risk. The risk is highest when adipose stores are depleted during migration, molting, or lactation. Transfer of these compounds through lipid-rich milk may further endanger neonates.

Reports following oil spills indicate that ingestion is not a major route of exposure for most pinnipeds. No oil was found in the digestive tracts of oiled grey seal pups (Davis and Anderson 1976), and petroleum hydrocarbon residues from the Santa Barbara spill in 1969 were not found in either blood or tissues of seals and sea lions (Simpson and Gilmartin 1970). Both seals and sea lions feed primarily on fish that live in the water column. Walruses (*Odobenus marinus*), however may be at greater risk since they feed on benthic infauna.

Ingestion of oil during grooming activity does not appear to be a major source of exposure in pinnipeds as it is in other species such as sea otters and polar bears (McLaren 1990).

Cetaceans

The order Cetacea comprises two suborders: mysticetes (baleen whales, of which there are eleven species) and odontocetes (toothed whales and dolphins, with sixty-eight species). These suborders are differentiated according to anatomical characteristics of mouth and feeding structures. Baleen whales feed by taking in large quantities of water and then filtering small invertebrates and fish as the water is expelled from between baleen plates suspended from the upper jaw. Odontocetes are toothed whales that feed primarily on fish and squid. Both anatomical characteristics and differences in feeding behaviors affect exposure risk of cetaceans to oil in the environment. Many species of both mysticetes and odontocetes inhabit U.S. coastal waters throughout the entire coastline.

Physical contact Impairment of thermoregulatory mechanisms is not a major risk in cetaceans since they depend primarily on a thick blubber layer for insulation and since oil does not readily adhere to their smooth skin. However, physical exposure to oil poses other risks. Fouling baleen plates with oil may decrease filtration capability and water flow (Geraci and St. Aubin 1982; Braithwaite, Aley, and Slater 1983). Species at greatest risk are those that feed at the sea surface (skim-feeding right whale, *Eubalaena glaciillis*, and bowhead whale, *Balaena mysticetus*, or lunge feeders such as humpback whale, *Megaptera novaengliae*) or at the bottom (gray whale, *Eschrichyitus robustus*). Furthermore, thick, weathered oil may also clog breathing passages.

The skin of cetaceans is a complex tissue with metabolic activity (Dargoltz, Romanenko, and Sokolov 1978; Geraci and St. Aubin 1982). To determine if exposure to oil causes pathological effects in skin of cetaceans, Geraci and St Aubin (1982, 1985) experimentally exposed four species of cetaceans, bottlenose dolphin (*Tursiops truncatus*), Risso's dolphin (*Grampus griseus*), whitesided dolphin (*Lagenorhynchus acutus*), and sperm whale (*Physeter catodon*) to crude oil and gasoline. Focal areas of skin were exposed for up to 75 minutes, and biopsy results showed that change varies with duration of exposure. Histological damage was reversible. Thus, cetacean skin appears to be more resistant to toxic effects from petroleum hydrocarbons than that observed in other mammalian species (Hansbrough et al. 1985).

Inhalation and Ingestion In a survey of fifteen species of cetaceans that were either stranded, caught in a fishery, or died in captivity, petroleum residues were identified in several tissues based on the presence of napthalenes (Geraci and St. Aubin 1982). Exposure history of these animals to petroleum is unknown. Though variance between individuals is significant, highest residue levels were in blubber (<1 to 25 ppm) and the liver (<1 ppm). The mixed-function oxidase system has been identified in cetaceans by the presence of cytochrome P-450 enzymes in liver (Geraci and St. Aubin 1982; Goksoyr 1986), suggesting the ability to metabolize petroleum hydrocarbons.

There is conflicting evidence about whether cetaceans can detect and avoid oil slicks, which would clearly affect their exposure risk. Experimental studies of captive, trained dolphins indicate that they avoided oil slicks after detection by visual, tactile, and echolocation means (Geraci and St. Aubin 1982; Smith, Geraci, and St. Aubin 1983; St. Aubin et al. 1985). Field observations, however, describe cetaceans swimming through and feeding in areas contaminated with oil (Goodale, Hyman, and Winn 1981; Geraci and St. Aubin 1982).

Mustelid

Sea otters are the only marine animal within the family Mustelidae, order Carnivora. Only one of two species occurs in U.S. waters; *Enhydra lutris* inhabits Pacific coastal waters from Alaska to southern California.

Physical contact Sea otters, smallest of all marine mammals, have only approximately 3 percent body fat (Williams et al 1988). To maintain body temperature, they depend on a dense fur coat and high metabolic rate, approximately 2.4 times that predicted in terrestrial mammals (Costa and Kooyman 1982). When oil destroys fur integrity, the insulating layer of air is lost and otters become hypothermic (Costa and Kooyman 1982). For this reason, sea otters are at high risk of morbidity and mortality when exposed to oil. Following the *Exxon Valdez* oil spill, over one thousand sea otter carcasses were recovered (Bayha and Kormendy 1990); this was considered only a fraction of the total mortality. Hypothermia and possibly asphyxiation were determined to be the most common causes of death.

Ingestion Contamination of fur leads to increased grooming behavior and inadvertent ingestion of oil (Williams 1978; Siniff et al 1982). Following an oil spill off Shetland, England, thirteen European otters (*Lutra lutra*) died. Necropsies performed on five of these animals determined that cause of death was hemorrhagic gastroenteropathy, likely associated with ingestion of oil (Baker et al. 1981).

Following the 1989 *Exxon Valdez* spill in Alaska, a major effort was undertaken to capture and rehabilitate sea otters exposed to oil. Of 357 sea otters captured, 123 died during treatment. A complete data set was available for 51 of these animals, so degree of oiling could be correlated with pathological change. Interstitial pulmonary emphysema, gastric erosion and hemorrhage, centrilobular hepatic necrosis, and hepatic and renal lipidosis were common in sea otters that had been contaminated with oil and absent or uncommon in uncontaminated otters (Lipscomb et al. 1993).

Sea otters feed primarily on macroinvertebrates (Estes et al. 1981), prey species known to accumulate petroleum hydrocarbon residues (Varanasi and Malins 1977; Neff 1979). Thus, sea otters are at risk of ingesting oil from both grooming behavior and contaminated food. Sea otters do not appear to consistently detect or avoid oil-covered waters either in the field or during experimental exposure in tanks (Baker et al. 1981; Siniff et al. 1982).

Manatee

The order Sirenia consists of two families, Trichechidae (manatees) and Dugongidae (dugongs). Only the West Indian manatee (*Tricheus manatus*) is found in U.S. waters; a population of approximately one thousand manatees inhabits warm coastal waters of Florida (Brownell, Ralls, and Reeves 1978), with occasional sightings from adjoining states. Like cetaceans, these animals are totally aquatic. Unlike any other marine mammal, manatees are strictly herbivorous, feeding on vegetation in the shallow coastal and riverine systems. Manatees are intolerant to cold temperatures, depending on warm water to maintain thermoregulation.

Very little information on potential effects of oil on these rare and unusual animals is available from the literature. There has been no major oil spill in Florida resulting in exposure of manatees. Deaths of dugongs in the Persian Gulf have been associated with oil both during the Iran–Iraq War in 1983 and again during the Persian Gulf War in 1991. Both times, several million gallons of oil were spilled into the marine environment, after which carcasses of dugongs and many other species were identified. Fifty-three dugong carcasses were found during the Iran–Iraq War ("Dugongs" 1983), and fourteen were found in the aftermath of the Persian Gulf War (Preen 1991). Dugongs were not examined during either event, however, and as a result, the specific effects of oil on these animals remain unknown.

By understanding the manatees' natural history and habitat requirements, it is possible to speculate on possible detrimental effects they may suffer in event of an oil spill. Though manatees live in the Gulf of Mexico, where extensive oil and gas exploration occurs, Bartz and Verinder (1980) determined that since wells were not located in coastal waters used by manatees, any spilled oil would likely dissipate before reaching manatee habitat and therefore would not pose a significant risk. If oil were spilled in manatee habitat, manatees would likely contact surface oil, which may irritate their eyes and mucous membranes or clog nostrils. Though direct thermoregulatory effects would not likely be significant, manatees may be susceptible to hypothermia if a spill were to displace them from their shallow, warm habitat out into deeper, colder waters. They may ingest oil if a spill were to wash into their feeding area and contaminate vegetation. Though spills may be localized and affect only a few animals, these are highly endangered animals, and it is necessary to protect each individual.

Polar Bears

The polar bear (*Ursus maritimus*) is the only marine species in the family Ursidae, order Carnivora. These Arctic inhabitants are thought to have evolved from the Siberian brown bear (Domico 1988). Polar bears are classified as marine mammals since they are dependent on the marine ecosystem for their survival. They live on pack ice and spend a considerable amount of time in the sea in search of prey, primarily seals. They are considered at risk to oil from both spills and blowouts associated with oil and gas exploration in the Arctic since they spend a great amount of time traveling across sea ice and traversing open water leads (Oritsland et al. 1981). Though accidental exposure has not yet occurred, experimental evidence suggests that polar bears are indeed sensitive to toxic effects from oil.

As part of the Eastern Arctic Marine Environmental Studies Program to determine acute and chronic effects of oil on polar bears, three subadult bears were exposed to a 1-centimeter-thick layer of crude oil on water in a tank for a period of 15 to 50 minutes (Oritsland et al. 1981). For 6 weeks following exposure, studies were conducted to characterize metabolism, thermal balance, health status, and grooming behavior. During this period. two of the animals died, the third survived.

During exposure, bears made no attempt to avoid contact with oil. Following fouling of their thick hair coat, the bears groomed intensively, ingesting oil during the process. All three bears showed clinical signs of shivering within hours, vomiting the following day, and diarrhea later. Polydipsia (increased thirst) continued throughout the remainder of the study period. Oil residues were identified in both vomitus and feces after immersion and in stomach contents and feces throughout the postoiling period (indicating continued ingestion from grooming). The urinary and hepatic/biliary systems appeared to be the main excretory means since urine and bile had high levels of oil residues after exposure and then lower levels throughout the study period.

Clinical expression of the toxic effects was latent, not becoming pronounced until 3 to 5 weeks after exposure. At this time, however, several systems showed significant change; acute anemia occurred in all bears and renal impairment was evidenced by dehydration and uremia. For the two bears that died, renal failure was the likely cause. Histological lesions were noted in brain, liver, gastrointestinal tract, bone marrow, lymphoid tissue, adrenal gland, lung, and skin. Furthermore, heat-balance studies established that oil exposure caused severe cold stress. From this study, it is clear that polar bears are at risk of adverse health effects from oil.

Sea Turtles

Five species of sea turtles inhabit U.S. coastal waters (National Research Council 1990). According to stranding data, abundance in decreasing order is loggerhead, Kemp's ridley, green turtle, leatherback, and hawksbill. Of these, loggerheads nest on the east coast of Florida, and some leatherbacks and green turtles nest on the Gulf Coast (National Research Council 1990).

As with marine mammals, adverse impacts of oil on turtles are poorly understood. Turtles can be exposed to oil on fouled beaches, where they nest; or at the sea surface, where they come to breathe, float, and feed. Physical coating with oil can retard mobility and impair vision of stranded turtles; oil droplets and tar balls are ingested by them (Balazs 1985; Gramentz 1986; 1988).

Following the June 1979 blowout of the Ixtoc 1 oil well in the Gulf of

Mexico, seven live, oil-contaminated turtles (six green turtles and one Kemp's ridley) and three carcasses (two green and one Kemp's ridley) were recovered by Hall and coworkers (1983). Two of three carcasses were emaciated, and though petroleum was identified in the gut, no evidence was found of associated gross or histologic lesions in the gastro-intestinal tract. No evidence of aspirated oil was identified in the lungs. Petroleum hydrocarbon residues were found in all tissues analyzed (kidney, liver, and muscle) from the three dead animals, and there was evidence of selective elimination of certain portions of the residues, suggesting chronic exposure. The authors suggest that prolonged exposure to oil may have weakened the animals, possibly by disrupting feeding behavior, and then exposure to a toxic component of oil or some other agent likely caused death.

Evidence of turtles' ability to detect or avoid oil slicks remains controversial. Gramentz (1988) reported that sea turtles did not avoid oil at sea, yet Vargo et al. (1986) reported that green turtles and loggerheads had a limited ability to avoid oil slicks under experimental conditions.

Loggerhead and green turtles exposed to weathered crude oil under laboratory conditions showed significant biological effects in several organ systems (Lutz, Lutcavage, and Caillouet 1989). Skin lesions included accelerated cell division, sloughing, and neoplastic response; lungs had reduced diffusion capacity and decreased oxygen consumption; gastrointestinal tracts showed decreased digestion efficiency; nares and eyelids had tissue damage; and there was evidence of change involving hematopoiesis, the immune ssystem, and salt glands.

Effects of oil on hatchability and development in two species of marine turtles was investigated both in the field (Kemp's ridley) and laboratory (loggerhead) by Fritts and McGehee (1981). They suggested that time of development during exposure and characteristics of oil would determine severity and type of effect. Contamination of nesting beaches prior to deposition of eggs may have little or no effect on reproductive success. Mortality occurred, however, when eggs halfway or more through incubation were exposed in the laboratory to oil-contaminated sand (30 ml oil/4 kg sand).

Thus as with marine mammals, sea turtles are vulnerable to adverse biological effects following exposure to oil in the marine environment.

Conclusions

Though marine mammals and sea turtles were not exposed to oil during the Arthur Kill spill, these species are clearly at risk from oil in the marine environment. When contingency plans are developed for oil spill response, it is critical that these animals be considered to ensure their protection. Further research is needed to better assess toxicological and biological effects caused by exposure to oil. Since all marine mammals and sea turtles are protected by law (Marine Mammal Protection Act of 1972 and Endangered Species Act of 1973), routine laboratory exposures are neither possible nor desirable. It is therefore important to maximize our utilization of any animals exposed to oil during catastrophic events such as oil spills. For example, following the *Exxon Valdez* spill, approximately 1,500 sea otters (both alive and dead) were examined by pathologists, allowing the opportunity to evaluate both sexes, all age classes, various degrees of oiling, and duration of exposure. Though we hope an exposure like this will never happen again, certainly another unexpected event involving oil will occur. It is important that appropriate protocols and plans be prepared in advance: A rapid response is essential.

We also need to improve our understanding of natural-history characteristics including geographical and temporal distributions and abundance of populations, habitat requirements, reproductive behavior, population status, and so on. This information will improve predictive capabilities to evaluate comparative risks under certain oil spill scenarios. As defined by damage assessment regulations in the Comprehensive Environmental Response, Compensation, and Liability Act (CERCLA) and the Oil Pollution Act (OPA) of 1990, an effort is currently underway to create a predictive model that will include much of this information (U.S. Department of the Interior 1991). Abundance and distribution of marine mammals, sea turtles, and birds by season is being evaluated in separate coastal regions of the United States.

Coordination and sharing of scientific information and joint contingency planning among federal and state agencies, academia, and the private sector are necessary both to improve our understanding of the biological effects of oil on marine mammals and sea turtles and to effectively respond to oil spills to maximize protection of these and all other species. To adequately be prepared for an oil spill, communities and state and federal agencies should

> have reliable data on the location and abundance of marine mammals and sea turtles within their jurisdiction,
>
> be able to quickly identify the areas of special concern that should be protected from the spreading oil, and
>
> have in place plans and protocols for rescue and rehabilitation efforts.

Literature Cited

Addison, R. F., and P. F. Brodie. 1984. Characterization of ethoxyresorufin O-de-ethylase in gray seal, *Halichoerus grypus*. *Comp. Biochem. Physiol.* 79C:261–263.

Addison, R. F., P. F. Brodie, A. Edwards, and M. C. Sadler. 1986. Mixed function oxidase activity in the harbour seal (*Phoca vitulina*) from Sable, N.S. *Comp. Biochem. Physiol.* 85C(1):121–124.

Baker, J. R., A. M. Jones, T. P. Jones, and H. C. Watson. 1981. Otter *Lutra lutra* L. mortality and marine oil pollution. *Biol. Cons.* 20:311–321.

Balazs, G. H. 1985. Impact of ocean debris on marine turtles: Entanglement and ingestion. In R. S. Shomura and H. O. Yoshida, eds., *Proceedings of the Workshop on the Fate and Impact of Marine Debris*, November 26–29, 1984, pp. 387–429. Washington, D.C.: U.S. Deptartment of Commerce, NOAA technical Memorandum NMFS. NOAA-TM—NMFS-SWFC-54.

Bartz, M. R., and S. H. Verinder. 1980. OCS oil and gas proposed 1981 sales A66 and 66: Draft environmental impact statement. Bureau of Land Management, New Orleans Outer Continental Shelf Office, New Orleans, La.

Bayha, K., and J. Kormendy. 1990. Sea Otter symposium. In Proceedings to evaluate the response effort after the T/V *Exxon Valdez* oil spill into Prince William Sound. *U. S. Fish and Wildlife Service Biol. Rep.* 90:1–485.

Braithwaite, L. F., M. G. Aley, and D. L. Slater. 1983. The effects of oil on the feeding mechanism of the bowhead whale. Final report to the Bureau of Land Management, U.S. Department of the Interior, Washington, D.C.

Brownell, R. L., K. Ralls, and R. Reeves. 1978. Report of the West Indian manatee workshop. In R. L. Brownell and K. Ralls, eds., *The West Indian Manatee in Florida*, pp. 3–16. Tallahassee, Fla.: Florida Department of Natural Resources, National Fish and Wildlife Laboratory, U.S. Fish and Wildlife Service, Sea World of Orlando.

Clark, R. C., Jr., and J. S. Finley. 1982. Occurence and impact of petroleum in the arctic environment. In L. Rey, ed., *The Arctic Ocean*, pp. 295–341. London: Macmillan.

Costa, D. P., and G. L. Kooyman. 1982. Oxygen consumption, thermoregulation and the effect of fur oiling and washing on the sea otter, *Enhydra lutris*. *Can. J. Zool.* 60:2761–2767.

Dargoltz, V. G., E. V. Romanenko, and V. E. Sokolov. 1978. Oxygen consumption by skin in dolphins and problem of cutaneous respiration in cetaceans. *Zool. Zh.* 57:768–776.

Davis, J. E., and S. S. Anderson. 1976. Effects of oil pollution on breeding grey seals. *Mar. Poll. Bul.* 7:115–118.

Domico, T. 1988. *Bears of the World.* New York and Oxford: Facts on File.

Dugongs, other marine life victims of Gulf oil spill. 1983 (July/August). *World Wildlife Fund News* 24:13–14.

Engelhardt, F. R. 1978. Petroleum hydrocarbons in arctic ringed seals, *Phoca hispida*, following experimental oil exposure. In *Proc. Conf. on Assessment of*

Ecological Impacts of Oil Spills, pp. 614–628. Keystone, Colo.: American Institute of Biological Sciences.

————. 1982. Hydrocarbon metabolism and cortisol balance in oil exposed ringed seals, *Phoca hispida. Comp. Biochem. Physiol.* 72C:133–136.

————. 1987. Assessment of the vulnerability of marine mammals to oil pollution. In J. Kuiper and W. J. Van den Brink, eds., *Fate and effects of oil in marine ecosystems*, pp. 101–115. Ordrecht, Boston, and Lancaster: Martinus Nijhoff.

Engelhardt, F. R., J. R. Geraci, and T. G. Smith. 1977. Uptake and clearance of petroleum hydrocarbons in the ringed seal, *Phoca hispida. J. Fish. Res. Board Canada* 34:1143–1147.

Estes, J. A., R. J. Jameson, and A. M. Johnson. 1981. Food selections and some foraging tactics of sea otters. In J. A Chapman and D. Pursley, eds., *Proc. of the Worlwide Furbearer Conference, 1980.* pp. 606–641, Froturg, Md.

Fritts, T. H., and M. A. McGehee. 1981. Effects of petroleum on the development and survival of marine turtle embryos. Washington, D.C.: U.S. Fish and Wildlife Service, Department of the Interior, contract no. 14-16-0.

Geraci, J. R., and D. J. St. Aubin. 1982. *Study of the Effects of Oil on Cetaceans.* Washington, D.C.: Bureau of Land Management, U.S. Dept of the Interior.

————. 1990. *Sea Mammals and oil: Confronting the Risks.* San Diego, Calif.: Academic Press.

————, eds. 1990. *Sea Mammals and Oil: Confronting the Risks. San Diego, Calif.: Academic Press.*

————. 1985. Study of the effects of oil on cetaceans. Final report to the U.S. Dept. of the Interior, Bureau of Land Management contract no. AA51-CTU-29, Washington, D.C.

Geraci, J. R., and T. G. Smith. 1976. Direct and indirect effects of oil on ringed seals (*Phoca hispida*) of the Beaufort Sea. *J. Fish. Res. Board Canada* 33:1976–1984.

Goksoyr, A. 1986. Initial characterization of the hepatic microsomal cytochrome P-450 system of the piked whale (Minke) *Balaenoptera acutorostrata. Mar. Environ. Res.* 19:185–203.

Goodale, D. R., M.A.M. Hyman, and H. E. Winn. 1981. Cetacean responses in association with *Regal Sword* oil spill. In E. K. Edel, M. A. Hyman, and M. F. Tyrell, eds., *A Characterization of Marine Mammals and Turtles in the Mid- and North Atlantic Areas of the US Outer Continental Shelf*, Cetacean and Turtle Assessment Program annual report. University of Rhode Island. Washington, D.C.: U.S. Department of Interior.

Gramentz, D. 1986. Cases of contamination of sea turtles with hydrocarbons. *U.N. Rocc. Info.* 17:25–27.

————. 1988. Involvement of loggerhead turtle with the plastic, metal and hydrocarbon pollution in the central Mediterranean. *Mar. Poll. Bul.* 19(1): 11–13.

Hall, R. J., A. A. Belisle, and L. Sileo. 1983. Residues of petroleum hydrocarbons in tissues of sea turtles exposed too the Ixtoc I oil spill. *J. Wildl. Dis.* 19(2):106–109.

Hansbrough, J. F., R. Zapata-Sirvent, W. Dominic, J. Sullivan, J. Boswick, and X. W. Wang. 1985. Hydrocarbon contact injuries. *J. Trauma* 25(3):250–252.

Hutchinson, J., and M. Simmonds. 1991. *A Review of the Effects of Pollution on Marine Turtles.* London: Greenpeace International.

Le Boeuf, B. J. 1971. Oil contamination and elephant seal mortality: A negative finding. *Biol. Bacteriol.* 1:277–285.

Lillie, H. 1954. Comments in discussion. In *Proc. Intern. Conf. Oil Pollution:* 31–33.

Lipscomb, T. P., R. K. Harris, R. B. Moeller, J. M. Pletcher, R. J. Haebler, and B. E. Ballachey. 1993. Histopathologic lesions associated with crude oil exposure in sea otters. Vet er. Pathology 30:1–11

Lutz, P. L., M. Lutcavage, and C. W. Caillouet. 1989. The effects of petroleum on sea turtles: Applicability to Kemp's ridley. In A. M. Landry, edr., Proceedings of the First International Symposium on Kemp's Ridley Sea Turtle Biology, Conservation and Management, October 1–4, 1985, Galveston, Texas, pp. 52–54. Texas A&M University Sea Grant Program.

McLaren, I. A. 1990. Pinnipeds and oil: Ecologic perspectives. In J. R. Geraci and D. J. St. Aubin, eds., *Sea Mammals and Oil: Confronting the Risks,* pp. 55–102. San Diego, Calif.: Academic Press.

National Academy of Sciences. 1983. *Oil in the Sea.* Washington, D.C.: National Academy Press.

National Oceanic and Atmospheric Administration. 1992. 15 C.S.R., Chapter IX, National Resource Damage Assessment, Notice of Proposed Rule Making. *Fed. Reg.* 57(50):8964–8988.

National Research Council 1983. *Oil in the Sea.* Washington D.C.: National Academy Press.

————. 1990. *Decline of the Sea Turtles: Causes and Prevention.* Washington, D.C.: National Academy Press.

Neff, J. M. 1979. *Polycyclic Aromatic Hydrocarbons in the Aquatic Environment: Sources, Fates, and Biological Effects.* London: Elsevier Applied Science Publishers.

Neff, J. M., and J. W. Anderson. 1981 *Response of Marine Animals to Petroleum and Specific Petroleum Hydrocarbons.* New York: Halstead Press.

Oritsland, N. A., F. R. Engelhardt, F. A. Juck, R. J. Hurst, and P. D. Watts. 1981. Effect of crude oil on polar bears. Environmental study report no. 24. Northern Affairs Program, Department of Indian Affairs and Northern Development. Ottawa, Ontario, Canada.

Preen, T. 1991. Gulf War oil spill. *Sirenews* 16:13–15.

Risebrough, R. W., W. Walker II, A. M. Springer, J. R. Clayton, E. F. Letterman, J. R. Payne, and T. T. Schmidt. 1978. A search for pollutants of petroleum origin in tissues of harbor seals, *Phoca vitulina,* in San Francisco Bay. Marine Mammal Commission, final report contract No. MA7AC007.

St. Aubin, D. J., J. R. Geraci, T. G. Smith, and T. G. Freisen. 1985. How do bottlenose dolphins, *Tursiops truncatus,* react to oil films under different light conditions? *Can. J. Fish. Aquat. Sci.* 42:430–436.

Simpson, J. G., and W. G. Gilmartin. 1970. An investigation of elephant seal and sea lion mortality on San Miguel Island. *Bioscience* 20(5):289.

Siniff, D. B., T. D. Williams, A. M. Johnson, and D. L. Garshelis. 1982. Experiments on the response of sea otters *Enhydra lutris* to oil contamination. *Biol. Conserv.* 261:272.

Smith, T. G., and J. R. Geraci. 1975. The effect of contact and ingestion of crude oil on ringed seals of the Beaufort Sea. In T. G. Smith and J. R. Geraci, eds., *Beaufort Sea Project.* Institute of Ocean Science technical report no. 5. Sidney, B.C.: Beafort Sea Project

Smith, T. G., J. R. Geraci, and D. J. St. Aubin. 1983. Reaction of bottlenose dolphins, *Tursiops truncatus,* to a controlled oil spill. *J. Fish. Res. Board Canada* 40:1522–1525.

U.S. Department of the Interior, 1991. 43 C.F.R., Part II, National Resource Damage Assessment, Notice of Proposed Rule Making. *Fed Reg.* 56(82): 19752–19773.

Varanasi, U., and D. C. Malins. 1977. Metabolism of petroleum hydrocarbons: Accumulation and biotransformation in marine organisms. In D. C. Malins, ed., *Effects of Petroleum on Arctic and Subarctic Environments and Organisms.* Vol. 2, *Biological Effects,* pp. 175–270. New York: Academic Press.

Vargo, S., P. Lutz, D. Odell, E. Van Vleet, and G. Bossart. 1986. *Effects of Oil on Marine Turtles: Vol. 1. Executive Summary. Vol. 2. Technical Report.* Florida Institute of Oceanography, Final Report MMS No. 14-12-0001-30063.

Warner, R. F. 1969. Environmental effects of oil pollution in Canada: An Evaluation of problems and research needs. *Can. Wild. Serv. Ms. Rep.* 645:16–17.

Williams, T. D. 1978. Chemical immobilization, baseline parameters and oil contamination in the sea otter. Marine Mammal Commission report no. MMC-77/06.

Williams, T. M., R. A. Kastelein, R. W. Davis, and J. A. Thomas. 1988. The effects of oil contamination and cleaning on sea otters (*Enhydra lutris*): I. Thermoregulatory implications based on pelt studies. *Can. J. Zool.* 66:2776–2781.

15. The Arthur Kill, People, and Oil Spills

Joanna Burger

New Jersey is highly industrialized and has the highest population density of all the fifty states. Despite the high level of industrialization, harbor facilities, and transportation arteries, even the most urban areas contain some open, wild places where some semblance of native ecosystems flourishes (Stansfield 1983). Such areas are important for a wide variety of recreational purposes, even when the places are small in area or are a narrow fringe along a river. The preservation of these natural areas often reflects a newfound pride in New Jersey by its residents (Pomper 1986) as well as the presence of abandoned land or intractable marshes and wetlands.

The Arthur Kill is a narrow tidal strait that separates Staten Island (New York) from New Jersey, a part of the Hudson–Raritan estuary consisting also of New York Bay, Kill van Kull, Newark Bay, East and Harlem rivers, and Raritan and Sandy Hook bays. To some people, the Arthur Kill is merely a highly industrialized complex of piers, refineries, and bulkheads. Indeed, the New Jersey shore of the Arthur Kill is 47 percent bulkheads (7% for New York's side), 6 percent riprap (1% for New York), 9 percent sand and gravel (19% for New York), and the rest mud flats and marshes (U.S. Army Corps of Engineers 1980).

To other people, however, the Arthur Kill is a ribbon of mud flats lined by marshes, with piers, bulkheads, and city parks for recreational activities; a waterway for fishing; and natural areas for wildlife (Burger, Parsons, and Gochfeld in press). Any degradation in the river itself can cause losses to the people who use the Arthur Kill for fishing and boating, birdwatching, and onshore activities such as walking, hiking, and other sports.

This chapter investigates how the public uses the Arthur Kill; examines how public use of the Kill was evaluated following the Exxon oil spill of January 1–2, 1990; reports on a study of fisherman and crabbers that use the Kill and discusses health concerns raised by their eating this catch; and suggests what information is necessary to evaluate public use of a waterway. The scope of this chapter is limited to noncommercial uses.

Framework for Human Use of a Waterway

People can use a waterway directly, either by engaging in activities on the water itself or along the shoreline, or they can participate in activities along the shore that are enhanced by the presence of the river but are not dependent upon it. These two categories include

Direct use	Indirect use
Boating	Hiking
Fishing	Walking and jogging
Crabbing	Sports (baseball, softball)
Clamming	Reading
Birdwatching	

An evaluation of the effect of degradation from an oil spill or other toxic accident must include the economic losses of both the direct and indirect uses. The loss of aesthetic values cannot be discounted if people no longer choose to walk, hike, or play along the Arthur Kill or any other waterway because of visual and olfactory degradation or health concerns.

The valuation of aesthetic and wildlife preferences is a new and developing field (Berryman 1987; Decker and Goff 1987; Fillion, Parker, and Dulvors 1988). Attention is devoted to examining social and economic values (Boyle and Bishop 1985; Brown and Manfredo 1987), measuring the economic value of wildlife and wild areas (Davis and Lim 1987), and using this knowledge to mitigate impacts (Stuckey et al. 1987). Although scientists are beginning to examine ways to value the use and importance of wild areas, wildlife, and natural ecosystems, the immediacy of pollution events and other disasters often militates against scientific research or input from an informed and concerned public, instead favoring quick corporate solutions and environmental-consultant reports. The latter are often out-of-town firms unfamiliar with the ecosystem under study. One objective of this chapter is to examine how natural resources and human use of these resources was evaluated during the Arthur Kill oil spill of 1990.

Public Use of the Arthur Kill

As a biologist working along the Arthur Kill for many years prior to the spill, I have observed a number of activities that require the use of the Kill directly, including boating, fishing, crabbing, and birdwatching. Indirect uses are also evident, primarily in parks that border the Arthur Kill. In travels along the Kill while engaging in research I always encountered people in small boats, birdwatchers, and fisherman, except in very cold or rainy weather. Although the Arthur Kill may appear degraded from the vantage point of the land, it appears less so from a small boat floating on the water. To children it is a wonderful playground with unlimited new things to discover.

I review here the work of Desvousges and Milliken (1991), who provided an economic assessment of damages from the oil spill. Their major findings were

1. No activities take place directly on the Arthur Kill and Kill van Kull. Instead, boating, water-skiing, fishing, and shellfishing take place in a number of areas adjacent to the Arthur Kill: Raritan Bay, Lower Bay, Sandy Hook, and the Atlantic Ocean.

2. The total number of slips available for recreation users to house their boats in the immediate Arthur Kill area is the sum of slips in four marinas (N=265 slips).

Desvousges and Milliken (1991) reached their conclusions from interviewing the owners of these four marinas. They assumed that operators would "know where their customers boated and fished and to what extent they used the Arthur Kill for these activities" (Desvousges and Milliken 1991: 3–6). From their telephone interviews of less than an hour each, they formed the above conclusions, as well as characterized recreational use on the Kill.

Their characterization included the fact that 100 percent of the marinas' customers used the Arthur Kill for *access* to other areas, most people were from New York or New Jersey; and that seasonality played a role in use of the marina facilities. The marina owners believed their customers felt that the Kill was not an aesthetically pleasing locale for recreation.

From these conclusions, Desvousges and Milliken (1991) computed a damage estimate for this boating access. They assumed that (1) the moored boats constituted 75 percent of the total boats in the area, (2) 10 percent of the boats were used each day, (3) 10 weeks were lost because of the oil spill, and (4) valuation for a boat was $30 a day. This leads to a total boat-access loss of $31,640 (Desvousges and Milliken 1991). Their best-case valuation was only $3,955, while their worst-case valuation was $221,480.

They used similar telephone interviews with three staff personnel from the Bayonne, Rahway, and Hudson Country Parks and Recreation Departments to examine indirect (or near-water) uses of the Arthur Kill. These interviews indicated that birdwatching, picnicking, softball, basketball, walking, and jogging were the primary recreational uses in the parks. From these and other interviews they used the percentage of user days affected (50%) over a 10-week period, and a valuation per day of $30 to compute a damage estimate of $14,250 for birdwatchers. The best-case damage assessment was $1,425, and the worst-case was $32,063 for birdwatchers (Desvousges and Milliken 1991). They did not conduct an assessment damage estimate for people who used the parks for picnics, jogging, walking, or sports because they could not determine the number of people involved or the frequency of use.

These economic assessments for the Arthur Kill are important because they establish economic losses for two user groups (access boaters and birdwatchers) and clearly indicate that people suffer directly from environmental degradation such as an oil spill. They are also enlightening because they provide a methodology for evaluating cost. In the case of the boaters, cost was largely based on a $30-per-day value, presumably the cost of boat rental. This rental fee, however, is unrealistically low (Safina, pers. comm.). Their estimates seem low even for some boaters because boat owners pay to rent their slips (which presumably they would stop doing if degradation from oil spills became severe and continuous); they spend money on fuel and maintenance for their boats (income lost for the marina owners); and they buy food, fish bait, and other supplies nearby for their outings. Similar costs attend birdwatchers.

However, I see a more fundamental problem with the approach of Desvousges and Milliken (1991), one that no doubt attends scientists unfamiliar with their area of study. The Arthur Kill flows through a highly industrialized and heavily populated area. Not all of the population is economically able to purchase a boat big enough for a slip, to pay for the slip, or to rent a boat for long-term use. Yet many residents along the Kill and its side creeks and channels have small rowboats tied to a stake, hauled up on shore, or moored to their own bulkhead. These people are unaccounted for in a survey of marinas. Many people fish and crab along the banks of the Kill yet are not "customers" of the marinas. Furthermore, people birdwatch and engage in other recreational activities at places other than parks along the Kill. Children often use the closest access, even when they pass through other people's yards or even the parking lots of industries. In essence, Desvousges and Milliken (1991) have examined the Kill along class lines. Unwittingly, their question, was, Of what value is the Kill to middle-class people? rather than Of what value is the Kill to all people who use it or even to local people?

On any warm spring or summer day a ride down the Kill will reveal these recreational users who do not use the Kill as an access waterway to somewhere else, but use the Kill directly. These small boats cannot easily reach Raritan Bay or the Atlantic Ocean. Along the banks and bulkheads fisherman and crabbers wait patiently. They are also unaccounted for in the survey of marina "customers." Children of all ages play on the marshes and mud flats, whether just looking for animals, fishing, or trying to pole small rowboats.

The results of Desvousges and Milliken's (1991) interviews are extremely important, but Desvousges and Milliken used only one method to evaluate use of a particular economic class. The other uses, which they overlooked, are more direct uses of the Kill itself. These users also may be more impacted by degradation due to an oil spill because they actually use the Kill, and do not merely boat through to more pristine places.

Fishermen and Crabbers on the Kill

The discrepancy between my own observations of people in rowboats, fisherfolk, and crabbers along the Kill, compared with Desvousges and Milliken's (1991) assertion that recreational activities along the Kill are limited or nonexistent, resulted in designing a study of fisherman on the Kill. From July through September 1991 John Brzorad, Neil Morganstein, and I interviewed forty-eight people who were crabbing or fishing along the Arthur Kill.

Our overall objective was to determine how fisherman and crabbers use the Kill, whether they eat their catch, and whether they perceived a health risk from their catch. The health hazards attendant to fishing in the Kill are largely unrelated to contamination from oil but relate to levels of heavy metals and PCBs (see Burger, Parsons, and Gochfeld in press). Nonetheless, these data can be used to indicate "nonmarina" use.

We interviewed fishermen and crabbers at marinas and along the banks of the Kill. We were dressed in our field gear, and we identified ourselves as being from Rutgers University. Before asking questions we admired their catch and asked what luck they were having. We then explained that we study the wildlife along the Kill and were also interested in how people use the Kill. We had a questionnaire and always asked questions the same way.

Most of the people interviewed were retired or blue-collar workers who lived nearby (Fig. 15.1). Many were alone (45%), but others had friends or relatives with them. They ranged in age from 11 to 77 years (average = 44, standard deviation [SD = 21]). They had been coming to fish or crab the Arthur Kill for an average of 11.4 years (SD = 15 years), although the range was from 2 or 3 months to 52 years.

Most of the people (87%) we interviewed were using cages or traps baited with chicken to catch blue crabs (*Callinectes sapidus*). They came to crab an average of 10.4 (SD = 10.4) times a month during the season

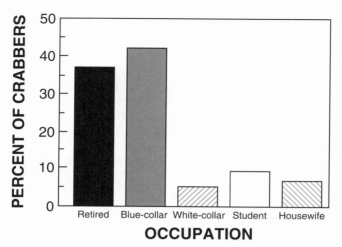

OCCUPATION

Figure 15.1. Composition of fishermen interviewed along the Arthur Kill.

(March to October) and said they ate an average of 9 (SD = 20.0) crabs per week. However, crabbers either seemed to eat fewer than 6 or more than 15 crabs a week (Fig. 15.2), depending on their ability to catch them. They reported that their spouses (usually women) ate 7.0 crabs (SD = 12.6) and their children ate 5.6 (SD = 11.7) crabs a week. Most people boiled the crabs, some fried them after they were boiled, and others used them in a soup. Many (62%) said they even ate the eyes.

At the end of each interview we asked them whether it was safe to eat the crabs: 20 percent said they had heard warnings, 17 percent believed they were not safe to eat, and 11 percent said pregnant women should not eat them. Yet when asked how many they ate each week only 8 percent said they did not eat them.

These results are interesting from three perspectives: (1) They indicate that people are directly using the Arthur Kill, (2) they raise the issue of whether crabbers are being exposed to toxics through ingestion of contaminated crabs, and (3) they raise the question of risk perception compared with actual risk. Clearly, these preliminary data indicate that people are using the Kill directly for fishing and crabbing and suggest that a detailed study of all uses is desirable. Such a study should also include children.

Both New Jersey and New York have areas closed to fishing and advisories on consumption of fish and shellfish in this area. New York State's Department of Health specifically advises eating no more than 6 crabs per week from the Arthur Kill. These regulations are based partly on PCB (polychlorinated biphenyl) levels as well as other toxics (Belton, Roundy,

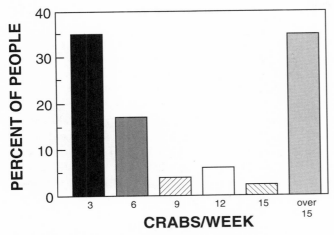

Figure 15.2. Crabs eaten by fisherman who use the Arthur Kill.

and Weinstein 1986). PCB's are responsible for a wide range of acute and chronic human-health effects including liver damage, reproductive disorders, and skin lesions (National Research Council 1979). They also cause multiple skin lesions characterized by chloracne (Fischbein 1991). Bioaccumulation in fish is a major route for introducing PCBs into the food chain, and eventually into people (Belton, Roundy, and Weinstein 1986; Squibb, O'Connor, and Kneip 1990).

From our interviews it is clear that many men, women, and children are consuming more than the recommended number of crabs. Of those interviewed, 47 percent said they ate more than 6 crabs a week. Some of the retirees said they ate as many as 124 a week and that crabs are a staple in their diet. These results indicate that the people who fish and crab along the Arthur Kill, largely blue-collar workers and retirees, are exposed to risks above the recommended level.

Belton et al. (1986) similarly reported that most fishermen in the Newark Bay and New York Bay areas ate their fish and that 90 percent ate their crabs. They conducted over one hundred formal interviews from 1983 to 1985 and observed over nineteen hundred people fishing. Many fishermen and crabbers used abandoned railroad trestles and piers rather than parks. Their findings were similar to ours in that many fishermen were retired people who returned time and time again to the same places, year after year. The older fishermen visited these sites 4 to 7 times a week (Belton, Roundy, and Weinstein 1986), similar to our average of 10.4 times a month. The fishermen resided in communities in the vicinity of the fishing and crabbing sites (Belton, Roundy, and Weinstein 1986).

They used the waterway not as access to somewhere else, but as a valued resource in itself.

Belton et al. (1986) also reported the excess cancer risks (i.e., above the general level in the population) from eating blue crabs from Newark and New York bays. The excess cancer risk from PCBs is 3.8 per 10,000 people who eat only 36.8 grams (1.2 oz.) per day of blue crab muscle (Belton, Roundy, and Weinstein 1986). If they eat the hepatopancreas with the muscle, the excess cancer risk jumps to 52 per 10,000. From our interviews on the Arthur Kill we found that almost 90 percent of the respondents ate both the muscle and hepatopancreas. The importance of communicating to the public the differences in risk based on the tissues consumed and the method of cooking cannot be underestimated (Ruppel and Sarner 1988), particularly for crabbers in the Arthur Kill.

They also noted that people did not perceive the risk from eating fish or crabs from the Kill, felt it was low because they ate few crabs, or believed it was only the water that was polluted. The difference between real risk, perceived risk, and actions is interesting in that perceived risk was low, yet risks from eating crabs might be high. This dichotomy in how people view risk has been the subject of recent research in risk perception in general (Slovic, Fischoff, and Lichtenstein 1979), and risk perception for fisherman in particular (Cordie, Locke, and Springer 1982; Burger and Gochfeld 1991). The bias to see a situation as risky yet not to see the risk as affecting oneself is called optimistic bias (Weinstein 1989). People often claim they are less likely to be affected than their peers. This "optimistic bias" appears with a number of risks such as food poisoning and radon (Weinstein, Klotch, and Sandman 1988).

The health risks from oil are less clear, although some oils (particularly recycled oil) contain heavy metals. The oil from the Exxon spill in January 1990 did not contain either PCBs or heavy metals. Edible fish and shellfish can accumulate oil or hydrocarbons in their tissue that remain for many months after exposure (Blumer et al. 1970). Presumably these residual oils would be ingested by people eating crabs and fish from an oil-contaminated area. The health effects of oil are difficult to determine because they usually occur in association with other toxics such as heavy metals and PCBs (Fischbein 1991).

For our purposes the observation that people are using the Kill directly for fishing and crabbing is pertinent. Although the Exxon spill occurred in January, when no one crabs or fishes, the oil was still on the mud flats and creeks in March (Engler 1990), when crabbers and fishermen returned to the Kills in full force. An oil spill during the crabbing and fishing season would more severely impact this user group.

It is difficult to determine the number of people using the Kill this way because although crabbers and fishermen concentrate along the piers and at the marinas (where we interviewed them), they also use any available access to the water. Since most of these users live nearby, their economic outlay in travel costs, equipment, and boat rental is small. Nonetheless, to the people who engage in these activities, the Kill is important for recreation and for food. Our interviewees would not part with their catch even when we offered them $2 a crab.

Another important way to view the consumption of fish and crabs by people is to consider humans as part of the food web. In this case people are top consumers, being at the top of the food web. As such, they eat crabs and fish that feed on organisms lower on the food chain, and that in turn feed on organisms still lower on the food chain. As toxics move up through the food chain, they accumulate in the tissues of each successive trophic level. Bioamplification is the process whereby species occupying higher trophic levels acquire higher concentrations of the toxics than organisms lower on the food chain (Hunter and Johnson 1982; Laskowski 1991). Often the concentration is an order of magnitude higher in each successive level of the food chain. Since people are at the highest level, they eat organisms that themselves have already accumulated high levels of toxics. Thus, contamination of any kind, whether it be from oil, PCBs, or heavy metals, has potential health effects on people who consume fish and shellfish from the waters of the Arthur Kill.

Conclusions

It is important to examine the overall damage assessment of the Arthur Kill oil spill of 1990 to evaluate the completeness and accuracy of the assessment, to identify important resources, and to provide useful data for future responses to catastrophic events. That Exxon commissioned a study of damages suffered by people attests to the importance of assessments in our present political and social climate. Acknowledging that recreational uses have value is another important step forward. Presumably such valuations will affect future policy decisions. The assessment itself was excellent for boaters who used the Kill in transit and for birdwatchers but failed to take into account a whole population that actually uses the Kill itself. This failing, I believe, comes about because the researchers were unfamiliar with the Kill and the surrounding communities and did not visit the area during warm weather, when fishermen and people in small boats were active. Experienced field scientists would identify the species of concern, sample the habitats of concern, conduct sampling over a representative time period, and sample the population adequately before drawing conclusions about behavior and use of an area.

At the very least, the Arthur Kill would be visited. For example, no ornithologist would merely interview (or observe as the case may be) a small segment of the population or fail to visit the site to observe the birds themselves. It seems crucial for people evaluating human use of an area to be at least as thorough as an ornithologist would be.

The evaluation of Desvousges and Milliken (1991) and our evaluation indicate that there are important direct and indirect uses of the Arthur Kill. But we suggest that an indepth study of human uses of the whole New York and Raritan Bay complex, including the Arthur Kill, would provide a much-needed baseline for the future.

Such a study should be conducted over 2 years to understand seasonal and yearly variations among regions and should involve regular on-site visits to quantify uses on and along the waterways. No ethnic or economic group should be excluded, and citizens should be involved in the study design (a lesson learned from the *Exxon Valdez* spill in Alaska) (Kelso and Brown 1991). Such a study would serve not only as a basis for understanding how the public uses the New York–New Jersey harbor estuary but would serve as a basis for evaluating the public health risk from the consumption of fish and shellfish. Although these activities are prohibited and restricted in many of these areas, the importance of these prohibitions is not reaching the using public. The influx of people from different cultures, who are often unfamiliar with the risks involved, will contribute to the problem in the foreseeable future.

Any urban area that is highly industrialized is vulnerable not only to massive catastrophic oil spills but to a series of small spills. The cumulative effect of years of small spills can have a devastating effect not only on the natural ecosystem, species diversity, and population numbers but on human uses. If the river is always just a little polluted with oil, the visual and olfactory effects may preclude use except by children and the most hardy of adults. Baseline data on human uses can help to demonstrate the importance of the available natural areas along the Arthur Kill, or any other aquatic area; change the political and social climate so these areas continue to be preserved and protected from future degradation; and lead to the improvement of overall environmental quality

Acknowledgments

I thank John Brzorad and Neil Morganstein for help with the interviews, Neil Morganstein and Michael Gochfeld for help with analysis, Michael Gochfeld, Caron Chess, and Carl Safina for comments on the chapter, and NIEHS (ES 05955) for partial support.

Literature Cited

Belton, T., R. Roundy, and N. Weinstein. 1986. Urban fishermen: Managing the risks of toxic exposure. *Environment* 28:19–37.

Berryman, J. H. 1987. Socioeconomic values of the wildlife resource: Are we really serious? In D. J. Decker and G. R. Goff, eds., *Valuing Wildlife: Economic and Social Perspectives*, Boulder, Colo.: Westview Press.

Blumer, M., G. Souza, and J. Sass. 1970. Hydrocarbon pollution of edible shellfish by an oil spill. *Marine Biology* 5:195–202.

Boyle, J., and R. C. Bishop. 1985. The total value of wildlife resources: Conceptual and empirical issues. In Associ. Environ. and Resources, economic workshop on recreation demand modeling, Boulder, Colo.

Brown, P. J., and M. J. Manfredo. 1987. Social values defined. In Decker and Goff, *Valuing Wildlife*, pp. 12–23.

Burger, J., and M. Gochfeld. 1991. Fishing a superfund site: Dissonance and risk perception of environmental hazards by fisherman in Puerto Rico. *Risk Analysis* 11:269–277.

Burger, J., K. Parsons and M. Gochfeld. In press. Avian populations and environmental degradation in an urban river: The kills of New York and New Jersey. In J. A. Jackson, ed., *Avian Conservation*, Madison: University of Wisconsin Press.

Cordie, F., R. Locke, and J. Springer. 1982. Risk assessment in a federal regulatory agency: An assessment of risk associated with the human consumption of some species of fish contaminated with polychlorinated biphenyls (PCBs). *Environmental Health Perspectives* 43:171–182.

Davis, R. K., and D. Lim. 1987. On measuring the economic value of wildlife. In Decker and Goff, Valuing Wildlife, pp. 65–75.

Decker, D. J., and G. R. Goff, eds. 1987. *Valuing Wildlife: Economic and Social Perspectives. Boulder, Colo.: Westview Press.*

Desvousges, W. H., and A. J. Milliken. 1991. An economic assessment of natural resources damages from the Arthur Kill oil spill. Report from Research Triangle Institute to Exxon Bayway Refinery, June 1991.

Engler, R. 1990 (March 5). Portrait of an oil spill. *Nations:*300:3020.

Fillion, F. L., S. Parker, and E. Dulvors. 1988. *The Importance of Wildlife to Canadians.* Ottawa, Canada: Canadian Wildlife Service.

Fischbein, A. 1991. Polychlorinated biphenyls (PCBs). In T. M. Last and R. B. Wallace, eds., *Public Health and Preventive Medicine*, pp. 440–458. Norwalk, Conn.: Appleton and Lange.

Hunter, B. A., and M. S. Johnson. 1982. Food chain relationships of copper and cadmium in contaminated grassland ecosystems. *Oikos* 38:108–117.

Kelso, D. D., and M. D. Brown. 1991. Policy lessons from Exxon *Valdez* spill. *Forum for Applied Research and Public Policy* 6:13–19.

Laskowski, R. 1991. Are the top carnivores endangered by heavy metal biomagnification? *Oikos* 60:387–390.

National Research Council. 1979. Polychlorinated biphenyls. Technical report to the National Academy of Sciences, Washington, D.C.

Pomper, G. M., ed. 1986. *The Political State of New Jersey.* New Brunswick, N.J.: Rutgers University Press.

Ruppel, B., and L. Sarner. 1988. Marine fish preparation guidelines. *New Jersey Fish and Wildlife Digest* 1:7.

Slovic, P., B. Fischoff, and S. Lichtenstein. 1979. Rating the risks. *Environment* 21:14–20.

Squibb, K. S., J. M. O'Connor, and T. J. Kneip. 1990 (October). New York/New Jersey harbor estuary toxics categorization. Draft report prepared for USEPA Region II.

Stansfield, C. A., Jr. 1983. *New Jersey.* Boulder, Colo.: Westview Press.

Stuckey, N. P., J. P. Bachant, G. T. Christoff, and W. H. Dieffenback. 1987. Public interest and environmental impact assessment and mitigation. In Decker and Goff, *Valuing Wildlife*, pp. 235–242.

U.S. Army Corps of Engineers (USACOE). 1980. EIS for Kill van Kull and Newark Bay Channels, New York and New Jersey. U.S. Army Corps of Engineers, New York district.

Weinstein, N. D. 1989. Optimistic biases about personal risks. *Science* 246:232–1233.

Weinstein, N. D., M. L. Klotch, and P. M. Sandman. 1988. Optimistic biases in public perceptions of the risk from radon. *Amer. J. Public Health* 78:796–800.

16. Ecological Risk, Risk Perception, and Harm: Lessons from the Arthur Kill

Michael Gochfeld and Joanna Burger

Initial reactions from some government officials, industrial representatives, and even some media following the 1990 oil spill in the Arthur Kill were that there were no significant natural resources or ecosystems to be damaged in what some viewed as a completely industrialized, urbanized, and degraded estuary. This chapter considers what was learned about resources at risk from the January 1–2, 1990, oil leak into the Arthur Kill and about risk perceptions. No doubt some people felt that if there had to be an oil spill this was an ideal place for it, and the kind of reaction normally elicited by fouled beaches and mud flats did not occur until the magnitude of the contamination became known and the importance of the natural resources was evaluated. One view succinctly expressed was

Twarn't nothing there before; tain't nothing thar now; why worry.
—Anonymous

Even when the public became outraged at the casual corporate response, a view validated by multiple subsequent oil leaks during the ensuing months, there was still little sense that a significant natural area had been damaged. Nonetheless, local conservationists recognized that significant habitats and wildlife existed, even in this degraded ecosystem, and these took on added importance when human recreational use was considered.

As the chapters in this book document, many natural features and a thriving salt marsh ecosystem can be found along both shores of the Arthur Kill. Plants, invertebrates, and vertebrates were harmed by the oil, and multiple human activities, such as food gathering and recreational uses, were impacted. In this chapter we examine the probability of comparable future oil spills, the potential risks to ecosystems, and the ways risk perceptions differ among the various parties.

Probability of Future Oil Spills

The history of major oil spills, particularly from grounded tankers such as the *Amoco Cadiz* and the *Exxon Valdez*, shows little change in technology or practices. Instead, the pattern of media and public outcry, delayed

Table 16.1. Major Oil Spills in New York Harbor and Environs Since 1942

Date	Location	Material	Volume (gallons)
Jan. 15, 1942	42 kms Long Island	Lubricant	3,500,000
Feb. 27, 1942	29 km E New Jersey	Fuel oil	3,900,000
March 10, 1942	18 km E New Jersey	Fuel oil	3,600,000
May 25, 1942	16 km E New Jersey	Crude oil	4,400,000
June 25, 1958	East River	Gasoline	300,000
Nov. 26, 1964	New York Harbor	Solvents	6,800,000
June 16, 1966	New York Harbor	Naphtha	5,900,000
May 22, 1970	New York Bay	No. 6 oil	100,000
Dec. 27, 1970	New York Bay	No. 2 oil	200,000
Jan. 23, 1972	New York Harbor	Crude oil	100,000
Feb. 5, 1977	Hudson River	No. 6 oil	480,000
Mar. 5, 1980	Arthur Kill/Raritan Bay	No. 6 oil	400,000
Jan. 1981	Kill van Kull	Crude oil	100,000
Feb. 1987	East River	No. 2 oil	301,000
Feb. 1987	Hudson River	Gasoline	102,000
Jan. 1, 1990	Arthur Kill	No. 2 oil	565,000
Feb. 28, 1990	Kill van Kull	No. 6 oil	30,000
March 6, 1990	Arthur Kill	No. 2 oil	127,000
June 7, 1990	Kill van Kull	No. 2 oil	260,000
July 18, 1990	Arthur Kill	No. 2 oil	40,000
Sept. 27, 1990	Kill van Kull	Waste oil	40,000

SOURCES: Gochfeld 1979; Natural Resources Defense Council 1990; newspapers 1990; Burger, Parsons, and Gochfeld in press.
NOTE: We define major oil spill as over 100,000 gallons. All of the 1990 spills over 25,000 gallons are listed.

cleanup, and slow ecosystem recovery, recurs with another event somewhere else. Although the preventive technology appears little changed, society has invested extensively in improved cleanup technologies. There is, however, no reason to assume that the risks of major spills are reduced by these technologies, although the risks can be partially mitigated.

Although ecologists voice concern about chronic and recurrent small-scale spills, pipeline leaks, and bilge cleanings, it is the catastrophic spills that serve to keep the issue of oil on troubled waters in the public's view. Table 16.1 lists documented major oil spills in the New York City area since 1942, and all spills over 25,000 gallons in 1990. It is likely that many small oil spills go unreported and that prior to 1970 only huge spills of over 1,000,000 gallons attracted attention. The number of huge spills has declined; however, medium-sized spills (100,000 to 1,000,000 gal.) continue to occur at the rate of about .5 per year, with four documented in the 1970s, four more in the 1980s, and three in 1990 alone. More than 50 percent occur in the Arthur Kill or Kill van Kull.

These data predict that without further controls, regulations, or modified safety procedures, significant spills, and occasionally even massive spills will continue to occur in the Arthur Kill and surrounding New York Harbor. Furthermore, these data suggest that other communities that have oil refineries, storage facilities, or transport routes are similarly at risk. Prudence calls for community awareness of risks and the development of both improved preventive strategies and improved response plans that include appropriate risk assessment and address perceptions of all parties.

Exxon Valdez and the Arthur Kill: A Case of Priming

Although conservationists argue (see Chapter 4) that the Arthur Kill spill got much less attention than it deserved, we suggest that it would have received even less attention if it had involved another company. From a risk perception viewpoint, we postulate a priming relationship between the 1989 *Exxon Valdez* spill in Alaska, which attracted more American attention than has any other oil spill, and the public response to Exxon's Arthur Kill spill. The hypothesis that public outcry was largely primed by the *Exxon Valdez* incident and by Exxon's response deserves testing with other case studies. We argue that if the Arthur Kill spill had happened to another company it likely would have attracted less notice.

One can also invoke an "outrage" hypothesis (Weinstein, Klotz, and Sandman 1988). Outrage rather than damage focused the public's attention on this spill. A corollary is that public response is not necessarily proportional to the magnitude of harm (Freudenberg 1988; Kasperson et al. 1988).

Definitions of Risk: Risk Assessment, Risk Management, Risk Perception

Risk is the probability of an adverse impact accompanying some event. It is the product of the probability of the event times the magnitude of the harm (Gochfeld 1991). It may be the risk of cancer in a population chronically exposed to a pesticide, or the risk of ecosystem degradation in an area acutely exposed to an oil spill. Damage, the actual measure of harm, can be assessed after an acute event but is usually not documented when the exposure is chronic and ongoing.

Risk assessment is the formalized, quasi-scientific process of estimating the likelihood of harm to some target population or system (Cohrssen and Covello 1989). It requires delineation of the target population and the biological endpoint(s) of concern, evaluation of the hazard and the exposure, and finally the quantitative characterization of the risk (National Research Council 1981, 1983, 1986). Thus far the process is supposed to be objective and value-free (Freudenberg 1988).

Risk perception is the individual or collective evaluation of the importance or magnitude of risk. It is analogous to risk assessment in that persons estimate the magnitude of risk, and one's perception may or may not be based on the results of risk assessment. However, people use a very different approach, which usually does not involve toxicological data or quantified exposure. Instead, risk perception includes the understanding of toxicity and one's own likelihood of exposure, modified by one's understanding of immediacy or seriousness of the hazard, its fearfulness, the voluntary or nonvoluntary nature of the exposure, and the costs of control or ability to control the risk (National Research Council 1989). Slovic (1987) demonstrated the importance of several of these subjective dimensions on one's perception of risk. It is apparent, and indeed inevitable, that different people or different parties will arrive at very different conclusions regarding risks because their perceptions are influenced by underlying differences in their personal experience, occupation, lifestyle, education, residence, and risk-taking behavior.

Risk management is the process of controlling risk, reducing its damage, and preventing its recurrence. It is a social–political–economic venture driven by risk assessment but modified by government and public perceptions and resources. Risk management can include legislation, regulation, and guidelines and policies imposed by government, as well as technological innovations, investments, remediations, and risk communications.

The costs of a particular technological change or policy that reduces risk enter strongly into risk management decisions, but the risk assessment process is intended to be free of cost–benefit considerations (National Research Council 1983, 1986). The fact that the costs of a catastrophe and the costs of catastrophe prevention are borne unequally by different parties is a most challenging feature of risk management.

Ecological Risk and the Arthur Kill

The field of risk assessment has evolved primarily in response to human health concerns, particularly cancer. In the past 5 years there have been formalistic attempts to develop ecological risk assessments and estimates of harm to ecosystems or particular components of these systems (O'Neill et al. 1982; Barnhouse 1992). This process involves identification of meaningful endpoints for the Arthur Kill.

The salt marsh ecosystem of the Arthur Kill contains, in addition to the aquatic and mud-flat substrate, a community of plants dominated by *Spartina alterniflora*, a biologically active interface between the tidal water and the marsh, and abundant invertebrate and vertebrate life that provides a food base for larger fish and ultimately colonially nesting waterbirds, wintering ducks, and some humans. An alternative to focusing on the ecosystem as a whole is to examine selected species as endpoints and to determine how their populations have fared. Both approaches are considered here.

The avian community of the Arthur Kill includes important breeding colonies of herons, egrets, and ibises. These are conspicuous birds that add an aesthetic dimension to the Kill and adjacent areas of Staten Island and New Jersey. These birds are valued by nature lovers and birdwatchers, and successful breeding colonies are essential to maintaining this aesthetic resource. Few people realize that the Arthur Kill colonies are among the largest in either New York or New Jersey (see Chapter 13).

In addition, two avian species of particular concern breed in the Kill. The yellow-crowned night heron (*Nyctanassa violacea*) is listed as endangered in New Jersey, and the northern harrier, or marsh hawk (*Circus cyaneus*), is endangered in New Jersey (NJDEPE 1992) and threatened in New York (NYDEC 1989).

Ecological Endpoints

The challenge for ecological-risk assessment in the Arthur Kill was to identify endpoints that could be quantified and to establish criteria for recognizing significant changes in these endpoints, and ultimately for estimating recovery. Although no formal risk assessment process was employed prior to the spill, postspill damage assessment did, de facto, identify endpoints of concern (see Chapters 2 and 3).

Although an ecosystem is more than the sum of its parts (Barnthouse 1992), there is no easy way to study large ecosystems, save by studying each component and its interaction with other components. Major syntheses have recently appeared (Bartell, Gardiner, and O'Neill 1992; Suter 1993) describing methods for approaching these complex tasks. Endpoints that can be used to evaluate the damage to the system must be identified (Sheehan et al. 1984; O'Connor and Dewling 1986; Suter 1990). A study can focus on the abiotic environment (soil, sediments, water), on the biotic components (plants, invertebrates, vertebrates), or on the different trophic levels (producers, consumers, decomposers). Although complex ecosystems may have hundreds of species at each trophic level, the salt marsh offers the opportunity to identify one or a few significant primary producers (Teal and Teal 1969), and often one can identify one or a few keystone predators whose activities influence the structure of the food web (Paine 1966).

Significant Impacts: Damage and Time to Recovery

Faced with the complete destruction of an ecosystem or the complete elimination of a species (particularly an aesthetically or economically important vertebrate), people can agree that a significant impact has occurred. But in many cases damage is partial, some members of most populations surviving to provide a nucleus for recovery of a system. There is no generally agreed-upon criteria for what percentage destruction is intolerable, nor is there consensus on how soon recovery must occur (or how close it must approach the original system) before it can be considered acceptable.

Resiliency, Recovery, and Risk Perception

Like a stretched spring that rebounds to its resting state, biological systems at all levels have certain homeostatic mechanisms that buffer them against change and tend to restore the system to its basal state after disturbance. Once overstretched, however, the spring loses its ability to recoil and remains damaged and unspringlike forever. A salt marsh completely covered and soaked by oil dies, and the oil-saturated mud may resist recolonization for decades. Whether, like the overstretched spring, it is ruined forever is uncertain. With lesser degrees of oiling (and depending on the type, conditions, and seasonality), the marsh vegetation and animals may be killed but recolonization may occur, even though traces of the oiling persist.

The rapidity of recovery is very variable, but Figure 16.1 illustrates several possible trajectories from rapid and complete recovery (a) or slow but complete recovery (b), to various incomplete recoveries (c, d, and e), to total failure (f). The recovery trajectory is a function of the intrinsic resiliency of the ecosystem—the effectiveness of its homeo-

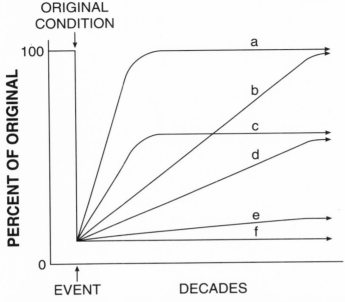

Figure 16.1. Alternative ecosystem recovery trajectories. Recovery can be measured in terms of population recovery, biodiversity indexes, similarly indexes, or some combination of these. All curves depict a massive destruction followed by (a) rapid total recovery, (b) slow total recovery, (c) rapid partial recovery, (d) slow partial recovery, (e) minimal recovery, and (f) no recovery.

static mechanisms—and the nature of the damage (including spatial extent and temporal features).

We postulate that if it is apparent that an ecosystem recovers quickly and completely, one should perceive a lower degree of risk, while slow, incomplete, or failed recovery should be associated with perceptions of greater risk. Although one may not know in advance what the recovery will be, as soon as the event occurs, scientists, managers, industry repesentatives, and the media must begin discussing the potential for recovery or for the consequences of long-term damage.

The fact that New Bedford Harbor, Massachusetts, showed signs of oil damage long after a major spill (Sanders et al. 1980) suggests that salt marshes may be slow to recover. On the other hand, the evidence from the Arthur Kill showed some signs of recovery even within the first growing season. We attribute this optimistic sign to the fact that oiling occurred in midwinter, when most organisms were naturally dormant and therefore least likely to be killed, and to the overall degree of oiling—the oil never reached the higher sections of the marsh.

Other Considerations

Health Risk

In addition to occupational exposures, oil products and the component petroleum hydrocarbons pose risks to human health from long-term, low-level exposure (MacFarland et al. 1984). This is mainly because some of the components have been identified as carcinogens, while others are neurotoxic. Nonetheless, human health is not usually considered important in evaluating the impact of oil spills. Most people are not exposed to oil during a spill, and special precautions are taken to minimize exposure of cleanup personnel (Schwope and Hoyle 1985). Furthermore, even when humans consume fish that have been contaminated with oil they are not likely to receive significant toxic doses of petroleum derivatives and carcinogenic polyaromatic hydrocarbons. The heavy metals and chlorinated hydrocarbons accumulated by the same fish from other sources no doubt pose a greater risk to human health.

Recreational Impacts

The Arthur Kill and particularly adjacent creeks and rivers are used recreationally by boaters and fishermen. Both activities were curtailed by the spill, but not as much as if the spill had occurred in the summer. The relative absence of large fish as evidenced by sampling in the summer of 1990 of fish that constitute prey for birds suggests that sports fish were also adversely affected (see Chapter 11). Remarkably, this sports' fish resource was ignored in Exxon's damage assessment because of a consultant's erroneous report that crabbing and fishing were not important in the impacted area (see Chapter 15). Moreover, the human population that uses the Kill in this way was relatively nonvocal, and its response to the impacts was not assessed. Thus, risk perception by the user public was never adequately evaluated.

Risk Perception and the Stakeholders: Lessons for the Future

People who deal with risk issues are aware that one's viewpoint and economic interests influence where one stands on any particular case and often influences one's view of risk assessment and risk perception in general. A party that causes acute or chronic pollution, a so-called responsible party, is likely to see much lower risks than an enforcement agency, which in turn sees lower risks than an outraged community does (Sandman 1986).

It is appropriate, therefore, to identify the stakeholders in any risk scenario: the local community, the general public, government agencies, responsible parties, and independent environmental groups. Based on many empirical cases we can array perceptions of risk from high to low as follows:

$$\text{local community} > \text{environmental group} > \text{general public} > \text{government} > \text{responsible party}$$

The Local Community

People who live, work, or play in close proximity to a contaminated site have an immediate concern of whether the pollutants have impacted or will impact their health or quality of life. They may move, stay indoors, avoid the polluted habitats, change their diet, or worry. Often one can identify a subset of users who specifically play, work on, or harvest resources from the affected site. These people are generally at highest risk from toxic exposures and are more likely to be adversely impacted by the disappearance of a resource or by advisories and restrictions on its use. Assuming that the exposure occurs involuntarily, the local community will experience the greatest outrage (Sandman 1986) and the user subset will experience the greatest impact. In the case of the Arthur Kill oil spill, people living along the Kill were impacted maximally, and those in surrounding communities that boat, fish, or crab on the Kill were also affected.

The General Public

The general public forms a continuum along the spectrum from people who can be categorized as risk averse (avoiding whatever risks they can at whatever cost they can tolerate), to risk seekers (such as hang gliders, bungee jumpers, and tobacco smokers). Thus, some people ignore risks while others exaggerate them. Slovic (1987) provides a "taxonomy" of risk, identifying those attributes that lead to a downplaying of risk. The converse is that there are certain subsets that ascribe fearsome attributes to certain hazards, while other people tolerate, ignore, or even enjoy the same. Bungee jumping and cigarette smoking are examples of activities that are viewed very differently by participants and nonparticipants.

The general public may overestimate risk by relying on the most fearsome data, choosing the really worst-case exposure scenarios, and assuming that all hazards act without threshold. They may also claim as intolerable the most sensitive endpoints, for example skin rash, or a modest and temporary population decline of some target species, even in the face of documented recovery potential. Or the public may underestimate risks that are subtle, remote, or nonimmediate.

For the Arthur Kill, the initial public response to the oil spill focused on the vulnerable egrets, herons, and ibises nesting on islands in the Arthur Kill because the public perceived these species as especially vulnerable. Gradually, public response expanded to include the food base (fish, crabs) for the birds and other vulnerable components of the ecosystem.

Government

Government is often accused by the public of underestimating risk (Sandman 1986) and by industry of overestimating risk (relying on worst-case exposure scenarios and upper 95% confidence limits). Freudenberg (1988) argues that scientists and technical experts are prone to overlook the importance of certain assumptions in adopting a risk-estimating methodology. Policy makers may be misled by very sophisticated risk estimation procedures and apparently precise numerical estimates, which hide magnitudes of uncertainty (Goldstein 1990).

Immediate government response following the Arthur Kill spill depended partially on knowledge and partially on perceptions of the resources at risk. Because of its stewardship of the Arthur Kill heronries, New York City's Department of Environmental Protection responded most quickly, whereas the state agencies' responses were not immediate.

Corporations

Corporations as a whole have acquired a reputation for underestimating the risks of their facilities, activities, or effluents, despite the fact that a number of large corporations have aggressively tried to reduce hazards and risks. Influenced by biases or vested interests, corporations tend to employ certain methodological mechanisms that lead to low estimates of risk. These include

1. Selection of only the most severe and incontrovertible endpoints

2. Focus on human endpoints to the exclusion of ecological ones

3. Focus on mortality rather than sublethal effects in nonhuman species

4. Reliance on human epidemiological studies to the exclusion of toxicological studies in animals or of bioassays

5. Selection of exposure scenarios employing only "realistic" as opposed to "worst cases"

6. Selection of nonconservative mathematical formulae and assumptions

All of these may be legitimate choices within the controversal bounds of how human-health risk assessment should be accomplished, but they tend to produce low estimates of risk.

There is an inevitable circularity in that a perception of low risk fosters risk assessment methodologies that calculate low risks, which in turn reinforces the original perception. In addition, scientists who perform risk calculations may be subject to pressures that impede a balanced report of data and risk (Schnaiberg 1980). Corporations have three risk perception problems: their own inherent viewpoints or biases related to economic concerns, the public's viewpoint or biases about industry, and the public's viewpoint or biases about a particular hazard itself.

Perceptions of the Impact of Oil Spills in General

In the media, oil spills are associated with slicks, oiled boats and beaches, and fouled or dead birds and mammals. This creates an aesthetic framework for viewing the risk from oil spills but does not take into account actual impacts on most trophic levels of the ecosystem. Scientists should also measure impacts on the producer and primary consumer trophic levels. These impacts may not be obvious or immediately important to any of the stakeholders, but they are important for an ecological-risk estimate.

Oil spills are one of the major catastrophes that capture media attention. Unlike natural disasters such as earthquakes and volcanoes, there is usually no loss of human life and there is always a human agency to blame, which makes such stories particularly attractive to the media and reading public. This human tendency explains why the public can be more concerned with hazardous waste than with naturally occurring radon, since there is no one to blame for the latter, even though it poses a much higher risk than the former (Weinstein, Klotz, and Sandman 1988).

The immediate impact of the oil—fouling of beaches and human property, and dead and dying wildlife—are conspicuous and persistent, keeping the story in the public eye for weeks. The impact on ecosystems, and in the case of both the *Exxon Valdez* and the Arthur Kill, litigation and study, keeps the story current for many months or years.

Various people in the risk arena marvel at how the "objective" impact and the public's risk perception ("subjective response") are often noncongruent. The apparent disparity results from the different values that each party brings to its perception of risk and from the technical person's failure to realize that what each party considers "objective" is partly just its own value-based perception of risk (Freudenberg 1988).

How does the actual, measurable impact of an oil spill influence the public perception of the risk (probability × harm) of oil spills? Does the public believe that the harm is of sufficient magnitude and the probability sufficiently high, that it is willing to bear the costs of, for example, requiring double-hulled oil tankers? Here it becomes important to distinguish a local and potentially impacted public from the general public.

A decade ago, in the aftermath of the Three Mile Island disaster, Slovic et al. (1980) showed that the lay public's perception of risks differed dramatically from that of scientific experts. The "public" ranked nuclear power as the most risky among thirty items, while "experts" ranked motor vehicles, handguns, smoking, and alcohol as the most risky items. The flaw in this approach is that the hazards associated with cars, guns, cigarettes, and alcohol are spread over all of society, while the risk from accidents at a nuclear or chemical plant are concentrated mainly in the vicinity of the facility, hence are borne unequally by society. Recent disasters at Chernobyl and Bhopal are a case in point (Shrivastava 1987).

Thus, we can postulate that a population in close proximity to a facility or spill should have a very different perception of the risk than one not directly affected. We can also argue that in addition to a spatial factor is a temporal factor. A population that has been primed by a recent catastrophe (even a nonlocal one such as the *Exxon Valdez*) will be more concerned and respond more vigorously to a recurrence of a similar event. Thus, oil spills beget concerns about oil spills. We postulate, however, that there is little spreading effect of one type of catastrope on sensitivity to another. Earthquakes beget concern about earthquakes but leave us unmoved about nuclear disasters or oil leaks, even though an earthquake affecting a power plant or refinery would be a major risk factor for both.

Reducing Risks from Oil Spills

It is reasonable to postulate that the magnitude of damage from an oil spill depends on its size, type, timing, and vulnerable targets and that the ecological risk associated with oil spills is influenced by these in conjunction with the frequency distribution of oil spills of certain sizes. If that is the case, one can develop plans for the prevention of oil spills and the minimization of damage from spills that do occur.

Primary prevention is aimed at stopping a problem before it even starts. This is the most desirable and effective approach to potential hazards and requires a combination of physical and administrative approaches. Secondary prevention is aimed at detecting a harmful event at the earliest possible stage so that it can be interdicted before harm occurs. Tertiary prevention involves recognizing and evaluating the damage and reversing its course, or rehabilitating the ecosystem to prevent long-term damage. This is clearly the least desirable and most expensive approach in the long run but has been the de facto approach used until the present.

Prevention of spills at stationary or mobile sources requires investment in planning, training, maintenance equipment (e.g., automatic shutoffs), double-hulled vessels, and improved transportation practices (e.g., containment of bilge cleaning water). Immediate and aggressive responses to spills at their earliest stage are important secondary-preventive steps.

Vulnerable ecosystems should be identified and have a baseline survey to characterize size, dominant species, and major pathways of energy flow and to identify indicator species that can be used to assess damage. Such inventories can be useful in determining how to proceed with cleanup if a spill does occur. Coastal ecosystems adjacent to petroleum shipping routes should receive high priority, and new approaches to computerized cataloguing and mapping of resources should be employed.

Emergency planning is essential. We propose that the amount of ecological damage is a nonlinear function of response time. The shorter the response time, the sooner a spill can be aborted and contained, or cleaned up. This reduces the volume of the spill and the size of the area requiring attention. The potential for damage increases, perhaps exponentially, with time and volume. Emergency planning must thus anticipate spills at any location and have the ability to mobilize a broad-spectrum response for any kind of petrochemical (or other chemical) spill.

Risk perception of chronic versus acute contamination is an important topic for community discussion and action. Although the record for preventing major oil spills is poor, that for preventing, controlling, or evaluating the impact of chronic spills is much poorer. The public rarely hears about such spills, and government agencies rarely attend to them. Many of the sources, for example, bilge washing outside territorial waters, are not easily regulated. Yet the total magnitude of such spillage and the extent of ecosystems affected may well exceed the damage from the dramatic spills. The overall perception of the risk from chronic oil spillage has not been formally studied but is probably disproportionately low compared with the impact of such spills.

The Role of Risk Assessment in Community Planning

Community planning and emergency planning are influenced by the risk perceptions of the stakeholders. Realistic perceptions of risk are those that accurately reflect the magnitude of risk, and often the public is remarkably good at estimating risk (Fischer et al. 1991). An accurate estimate certainly does not imply that all stakeholders will form the same perceptions, however. Perceptions are shaped by values as much as by facts. And values in turn are shaped by economic self-interest, social outrage, environmental awareness, and aesthetics, as well as the probabilities that the event will occur and that damage will follow.

Contingency valuation of ecological resources is emerging as a new and contentious field of law, economics, and risk management. Only by understanding both the risk and the risk perceptions of all stakeholders can future planning for catastrophes such as oil spills be effective.

Acknowledgments

Over the years we have profited from discussions with many people about risk; we thank particularly Caron Chess, Michael Gallo, Bernard Goldstein, Michael Greenberg, and Daniel Wartenberg. Our studies have partially been funded by NIEHS Grant ES 05022 to the Environmental and Occupational Health Sciences Institute and NIEHS ES 05022.

Literature Cited

Barnthouse, L. W. 1992. Case studies in ecological risk assessment. *Environ Sci Technol* 26:230–231.

Bartell, S. M., R. H. Gardiner, and R. V. O'Neill. 1992. *Ecological Risk Estimation*. Boca Raton, Fla.: Lewis.

Burger, J., K. Parsons, and M. Gocheld. In press. Avian populations and environmental degradation in an urban river: The kills of New York and New Jersey. In J. A. Jackson, ed., *Avian Conservation*. Madison: University of Wisconsin Press.

Cohrssen, J. J., and V. T. Covello. 1989. *Risk Analysis: A Guide to Principles and Methods for Analyzing Health and Environmental Risks*. Washington D.C.: Council on Environmental Policy.

Covello, V. T., J. Menkes, and J. L. Mumpower, eds. *1986 Risk Evaluation and Management*. New York: Plenum.

Douglas, M., and A. Wildavsky. 1982. *Risk and Culture*. Berkeley: University of California Press.

Fischer, G. W., M. G. Morgan, B. Fishchoff, I. Nair, L. B. Lave. 1991. What risks are people concerned about? *Risk Analysis* 11:303–314.

Freudenberg, W. R. 1988. Perceived risk, real risk: Social science and the art of probabilistic risk assessment. *Science* 242:44–49.

Gochfeld, M. 1991. Risk assessment. In J. Last, ed., *Maxcy-Rosenau-Last Public Health and Preventive Medicine*, 13th ed., pp. 332–342. Norwalk, Conn.: Appleton-Century-Crofts.

Goldstein, B. D. 1990. The problem with the margin of safety: Toward the concept of protection. *Risk Analysis* 10:7–10.

Kasperson, R. E., O. Renn, P. Slovic, H. S. Brown, J. Emel, R. Globe, J. X. Kasperson, S. Ratick. 1988. The social amplification of risk: A conceptual framework. *Risk Analysis* 8:177–187.

MacFarland, H. N., C. E. Holdsworth, J. A. MacGregor, R. W. Call, and M. L. Lane. 1984. *Applied Toxicology of Petroleum Hydrocarbons. Advances in Modern Environmental Toxicology*, vol. 6. Princeton, N.J.: Princeton Scientific Publishers.

National Research Council. 1981. *Testing for Effects of Chemicals on Ecosystems*. Washingtonn, D.C.: National Academy Press.

———. 1983. *Risk Assessment in the Federal Government: Managing the process*. Washington, D.C.: National Academy Press.

———. 1986. *Ecological Knowledge and Environmental Problem-Solving*. Washington, D.C.: National Academy Press.

———. 1989. *Improving Risk Communication*. Washington D.C.: National Research Council, Committee on Risk Perception and Communication, National Academy Press.

New Jersey Department of Environmental Protection and Energy (NJDEPE). 1992. *Special Animals of New Jersey*. Trenton: New Jersey Department of Environmental Protection and Energy.

New York Department of Environmental Conversation (NYDEC). 1989. *Endangered Species in New York*. Albany: New York Department of Environmental Conservation.

O'Connor, J. S., R. T. Dewling. 1986. Indices of marine degradation: Their utility. *Environ. Manage.* 10:335–343.

O'Neill, R. V, R. H. Gardner, L. W. Barnthouse, G. W. Suter II, S. G. Hildebrand, and G. W. Gehrs. 1982. Ecosystem risk analysis: A new methodology. *Environ. Toxicol Chemistry* 1:167–177.

Paine, R. T. 1966. Food web complexity and species diversity. *American Naturalist* 100:65–75.

Sanders, H. L., J. F. Grassle, G. R. Hampson, L. S. Morse, S. Garner-Price, and C. C. Jones. 1980. Anatomy of an oil spill: Long term effects from the grounding of the barge *Florida* off west Falmouth, Massachusetts. *J. Marine Res.* 38:265–380.

Sandman, P. 1986. *Explaining Environmental Risk*. Washington, D.C.: Office of Toxic Substances, U.S. Environmental Protection Agency.

Schnaiberg, A. 1980. *The Environment: From Surplus to Scarcity*. New York: Oxford University Press.

Schwope, A. D, and E. R. Hoyle. 1985. Personal protective equipment. In S. P. Levine and W. F. Martin, eds., *Protecting Personnel at Hazardous Waste Sites*, pp. 183–214. Boston: Butterworth.

Sheehan, P. J., D. R. Miller, D. R. Butler, and P. Bourdeau, eds. 1984. *Effects of Pollutants at the Ecosystem Level.* Chichester, England: Wiley.

Shrivastava, P. 1987. *Bhopal: Anatomy of a Crisis.* Cambridge, England: Ballinger.

Slovic, P. 1987. Perception of risk. *Science* 236:280–285.

Slovic, P., B. Fischoff, S. Lichtenstein. 1980. Facts vs. fears: Understanding perceived risk. In R. Schwing and W. A. Albers, Jr., eds., *Societal Risk Assessment: How Safe Is Safe Enough?* pp. 181–216. New York: Plenum.

Suter, G. W. II 1990. Endpoints for regional ecological risk assessment. *Environ Manage* 14:9–23.

———. 1993 *Ecological Risk Assessment* Boca Raton, Fla.: Lewis.

Teal, J. M., and M. Teal 1969. *Life and Death of a Salt Marsh.* Boston: Little, Brown.

Weinstein, N. D., M. L. Klotz, P. M. Sandman. 1988. Optimistic biases in public perceptions of the risk from radon. *Amer. J. Public Health* 78:796–800.

Conclusions

17. From the Past to the Future: Conclusions from the Arthur Kill

Joanna Burger

Wetlands, including tidal bays and estuaries, are critical habitats for many species of plants and animals. Estimates of wetland loss since presettlement days of North America range as high of 55 percent. Coastal bays and estuaries are particularly vulnerable because a large percentage of the world's population is concentrated along coasts, a fact that is true of the United States as well. Wherever there are large concentrations of people, the environment suffers from development, habitat loss, and pollution.

The Arthur Kill and associated Kill van Kull, Newark Bay, and Hackensack River, pass through one of the most highly urbanized and industrialized areas of the world. Although the banks have in large parts been bulkheaded, many salt marshes remain, providing habitat for the complex food web that culminates in the herons, egrets, and ibises that breed on islands in the Arthur Kill. The problems the Kill faces are those of any urban river: sewage pollution, chemical and oil pollution, boat traffic, garbage, channelization, industrial effluent, habitat destruction, and human disturbance. Nonetheless, the fact remains that the Kill supports functioning salt marsh ecosystems with a wide variety of invertebrate and vertebrate species living there.

The single most important characteristic of the Arthur Kill that determines how it is viewed is the contrast between heavy human development and a functioning natural ecosystem. Many who see the Kill see only the industrial complex; bulkheads; old, abandoned cars and other garbage littering the marshes and creeks; and a film of oil over the water and marsh. This is how it appears from the shore. From a boat, however, the Kill is bordered by mud flats dotted with white herons, dark ibises, and foraging gulls.

The January Oil Spill

When the massive Exxon oil spill of 1990 occurred, massive for a river only 25 kilometers (15 mi.) long, many people (including some government officials and Exxon representatives) discounted the potential for damage, because the area was already polluted and few natural resources remained. Initially the hue and cry came only from conservationists and scientists, followed almost immediately by the New York City Department of Environmental Protection. The responsible state agencies did respond once the magnitude of the potentially impacted resources became known.

Because the oil spill occurred in mid winter, the immediate effects on plants and wildlife were less severe than they would have been had it occurred in the summer: only because many species were dormant (plants) or hibernating or inactive below the mud (terrapins, many invertebrates), or because they had already migrated from the area (most birds). Unfortunately, sufficient oil spread on the mud flats to kill outright innumerable invertebrates that lived on the surface, birds and mammals that were covered in oil, and fiddler crabs (*Uca* spp.) and terrapins (*Malaclemys terrapene*) that were forced from their overwintering sites below the mud along tidal creeks and marshes.

Cleanup operations included placing booms to prevent oil from entering salt marsh creeks and from landing on Pralls Island (with its sensitive heronry habitat), using mechanical skimmers to remove oil from the water surface, and retrieving oiled wildlife for toxicological assays and rehabilitation. Unfortunately, several hours passed between the rupture of the Exxon oil pipeline and the placement of booms. In the meantime, the oil entered creeks and landed on Pralls Island. Thus, in some cases the booms merely kept oil in the creeks rather than allowing natural tidal flow to wash it away, to disperse over the larger Raritan and Newark bays. Nonetheless, the periodic changing of the absorbent booms did remove large quantities of oil.

The Following Years

In the aftermath of the spill representatives from several government agencies (local, state, and federal) and conservation groups, scientists, and concerned citizens met to plan the cleanup, oversee the damage assessment, and determine appropriate mitigation and restoration. Because of the impossibility of assessing the effects of the oil spill on the whole ecosystem, indicator species were selected for monitoring by the coalition, composed of representatives from New York and New Jersey state agencies, the New York City Department of Environmental Protection, and adjacent towns. Since the nesting colonies of herons, egrets, ibises, and gulls are a unique and important resource in the Arthur Kill, they formed the focus for the damage assessment. Thus, in the year following the oil spill there was a detailed evaluation of population levels, foraging behavior, and reproductive success of these colonially nesting waterbirds. The damage assessment also included evaluating population levels and behavior of the shrimp, fiddler crabs, and fish that are the principal foods for the waterbirds. Presumably, changes in the prey base would seriously affect reproductive success and population levels of the herons, egrets, ibises, and gulls. These species thus served as a set of indicators that represent the complex system. They represent several trophic levels, including top carnivores.

Other studies, not directly part of the damage assessment, were conducted on foraging behavior of waterbirds, nesting behavior of gulls, immediate behavior of fiddler crabs, cancerous lesions of clams (*Mya arenaria*) and blue crabs (*Callinectes sapidus*), and vegetation growth (all discussed in this book).

In the first year following the spill nearly 20 percent of the *Spartina* fringing the waterway was dead or still showed effects from the oil. The vegetation most affected was adjacent to the water–land interface where high tides had washed the oil immediately after the spill. Unfortunately, this is also the zone of highest invertebrate use, and the zone where the waterbirds forage. In the second and third years following the spill, some areas showed recovery, with revegetation of underground stems growing from adjacent, healthy *Spartina*. Some areas, however, still showed no vegetation recovery, and a wide swath of mud flat was still exposed above mean high tide.

Invertebrates living in the intertidal zone, for example, clams, blue crabs, and fiddler crabs, showed immediate behavioral effects and mortality following the spill. From 4 to 6 months after the spill, fiddler crab numbers were down immediately across from the spill, and behavioral differences that might impact their survival and reproduction were evident. In 1991, 1992, and 1993 these differences continued to decrease, although they have not yet completely disappeared.

Clams living in the Arthur Kill generally show a higher incidence of lesions in several internal organ systems when compared with clams living elsewhere in New Jersey. Nonetheless, following the rupture of the Exxon pipeline in 1990, the incidence of moderate to severe lesions of the digestive gland, intestine, and mantle increased rapidly. There were other changes as well. The intestinal lesions persisted for about 4 months after the spill. The incidence of lesions, however, returned to prespill levels by 1991 and 1992. Thus the clams, at least, demonstrate the rapid ability of some organisms to recover from an oil spill.

Some size classes of fish living in the Arthur Kill (i.e., mummichogs, *Fundulus heteroclitis*) were less abundant for 2 years following the oil spill of 1990 than they had been the previous year. Smaller-sized mummichogs decreased in the year following the oil spill but increased in 1991. Grass shrimp (*Palaeomonetes* spp.) declined in numbers after the oil spills of 1990 and have not yet recovered even 2 years later.

Birds are perhaps the most readily impacted by oil spills. Since the spill occurred in the winter, the loss of insulative properties of the feathers had devastating effects. Oiled birds could not remain warm. Despite their ability to move from the environment immediately follow-ing a spill, birds that bred in the Kill the following summer could not abandon their eggs or young. The heronries in the Arthur Kill are an important aspect of the local ecosystem and are critical components of the ardeid communities in both New York and New Jersey. Species feeding in tidal habitats experienced lowered reproductive success and disrupted foraging ecology in the breeding season following the 1990 oil spills. Most losses were attributed to starvation of the young due to habitat degradation and diminished prey abundance.

In 1991 and 1992 changes in the colony site use, species abundance, and reproductive rate of some species were still evident. Snowy egrets, the bird species most tied to the tidal marshes, (*Egretta thula*) continued to have lowered reproductive success and a decreased breeding popula-tion. Only by 1993 did reproductive success reach prespill levels.

Risk, Risk Perception, and Risk Management

Clearly the risk of an oil spill in the Arthur Kill and surrounding bays, rivers, and estuaries is high, both in terms of past spill history and with respect to the concentration of oil storage and transfer facilities in the New York–New Jersey harbor estuary. Furthermore, the daily oil tanker traffic and number of underground pipelines (getting older every day) increases the likelihood of an accident.

Not only is the Arthur Kill exposed to the potential of acute oil spills, but it is also vulnerable to chronic, low-level spills that, in total, could far exceed one large oil spill. The effect of chronic spills is usually ignored by the public and the media even though the total oil spilled from chronic exposure may be substantial. Chronic oil spills are also problem-atic because the ecosystem is never free from oil for long enough to recover adequately before another major spill occurs.

Risk, however, involves not only the probability of an event but the magnitude of the damage. Clearly, high concentrations of oil can have obvious and immediate lethal effects on a variety of plants, invertebrates, and vertebrates. This happened following the January 1990 oil spill. The damage was less than it would have been if the spill had occurred during the spring or summer, but many organisms died nonetheless.

Furthermore, all components of the Arthur Kill ecosystem that were examined in the spring and summer months following the oil spill showed some damage in either direct mortality (*Spartina*, large prey fish, bait fish crabs), behavior (fiddler crabs, fish), lesions (clams, bait fish), or nest-site selection (gulls), and reproductive success (egrets).

In the 2 years (1991 and 1992) following the initial spill year some recovery was evident in all components of the affected Arthur Kill ecosystem. In some areas *Spartina* vegetation moved in, fiddler crabs showed less-severe behavioral abnormalities, clams returned to ambient levels of lesions, fish populations returned to prespill levels, gulls began to exhibit prespill habitat selection, and some of the initial reproductive effects in the waterbirds were less extensive. Some species, such as snowy egrets, continued to show severe reproductive effects even 3 years later (1992). The results from our combined studies on the Arthur Kill following the oil spill indicate that such a massive spill has immediate effects and that these effects are most severe in the year following the spill. In the absence of another similar spill, the organisms in the Arthur Kill have begun to recover, and some have already recovered to prespill levels. This not only provides some hope for the future of the Arthur Kill ecosystem but also illustrates the recuperative properties of natural systems.

Lessons for the Future

The Arthur Kill oil spill of 1990 was not an isolated event, either for the New York–New Jersey harbor estuary or for coastal areas in general. Oil spills occur all the time: What varies is the magnitude of the spill, the extent of the damage, and the completeness of media coverage. What also varies is the degree of preparedness of the communities affected.

The major lesson we learned from the Arthur Kill is the importance of being prepared. In Table 17.1 I outline some of the issues that must be thought out for a community, state, region, or country to be prepared for an oil spill. Although the list is short, the tasks outlined are by no means easily arranged or quickly accomplished. Nonetheless, the key to being able to respond immediately to an oil spill includes identification of natural and human resources at risk; an inventory of potential hazards from local industries or transport facilities; an inventory of endpoints for assessment should there be a spill; an inventory of response personnel and equipment; an established list of government agents, industry representatives, scientists, and conservationists who could respond to both the immediate spill and the need for a damage assessment; and a well-developed response procedure.

Communities would also be well advised to identify scientists who can develop restoration plans for damaged ecosystems. Restoration ecology

Table 17.1. Protocol for Environmental Catastrophe Preparedness

I. Identify resources at risk
    1. Natural resources
        a. Vulnerable ecosystem
        b. Endangered and threatened species
        c. Species of concern (populations that are near threatened status)
        d. Species or resources of special importance or aesthetic value
            (such as the Arthur Kill heronries)
    2. Human resources
        a. Recreational and aesthetic
        b. Food (crabbing, fishing)
        c. Local businesses dependent on resources (marinas, bait shops)
II. Inventory local hazards
    1. Industrial and agricultural
    2. Storage and transport facilities
III. Inventory emergency response and cleanup personnel, and necessary
      equipment
IV. Identify potential natural- and human-resource endpoints for monitoring
     and assessment
V. Establish liaison among stakeholders
    1. Government
        a. Federal
        b. State
        c. Local
    2. Industry
    3. Scientists
        a. Natural
        b. Social
        c. Risk analysts
        d. Rehabilitators
    4. Conservationists
    5. Public
        a. Local/users
        b. Nonlocal
VI. Educate media about local resources, hazards, and risks
VII. Develop response procedures, designate levels of decision making, and
      open lines of communication

is an emerging discipline that will grow in importance as we try to restore damaged habitats to viable, functioning ecosystems. As it becomes more evident that humans and other components of the ecosystem must live in balance or we will not have sufficient resources to live happy, healthy, aesthetically pleasing lives, scientists and conservationists who are able to mitigate the damage from human-induced perturbations or catastrophes will become more important.

It is evident that we must learn how to translate the lessons learned from one type of perturbation (such as an oil spill) to others (such as toxic-chemical spills). The list of preparations and data necessary to cope with an oil spill (Table 17.1) can easily be used for any environmental pollutant. Knowing the natural and human resources at risk, as well as the local hazards, will prove useful for any catastrophe. Understanding the natural and human resources in an area is also important for other aspects of human endeavor such as development and the creation of parks, reserves, and green places.

It is also evident that we must learn to value ecosystems from the perspective of the organisms living there rather than from our own values of what is aesthetically pleasing. Although we may see only abandoned cars, tires, garbage, oil, and other remnants of degradation, the fact that the Arthur Kill holds one of the largest viable heronries in the northeastern United States indicates that the ecosystem of the Arthur Kill is, in the most important senses, functioning. We must learn to use the resident organisms as indicators of environmental health and seek to continue improving their environment as well as our own.

18. Epilogue: A Matter of Viewpoint

Joanna Burger

A large, sweeping span of silver steel glistens in the sunlight, forming a gentle arch from one green shore to another. Sitting amidst a *Spartina* salt marsh, surrounded by gently waving grass, I watch the tidal waters slowly receding along glistening gray mud. The bridge merely frames the marshes and river, rising majestically above the tidal swales.

Spartina grows thick and lush, the dark green stems sticking stiffly above the mud, forming a wide sea of grass that lines the river, bordering the meandering creeks, stretching from the river banks to the land. Back on firmer ground, where plant species diversity increases, the soft sameness of *Spartina* gives way to a myriad of native and exotic wildflowers and weeds. The mud banks are dotted with tiny holes less than a half inch wide, each separated by a few inches from its neighbor. Slowly exposed by the receding tide, the hole is hidden by a film of water. Suddenly the water implodes, revealing a dark tunnel. Minutes pass, and the openings no longer are obvious, shadows seem to fill the holes. Slowly, tiny legs move outward, and then a small body appears, eyes extending upward on slender stalks. Suddenly over the whole bank abrupt movement catches the sunlight. Fiddler crabs, some moving slowly while others dash short distances, begin the task of defending territories, finding food, seeking mates. Using their two-pronged cheliped, they shove algae and detritus into their mouths and probe the mud for yet more food. With their one large and ungainly cheliped, males seem at a disadvantage, for they can search for food only with their small cheliped while females probe and shovel in food with both at once. Males use the large cheliped for territory defense and for attracting females.

Only a trickle of water flows slowly in the lowest reaches of the creek. The dark ribbed mussels peeking above the mud are closed, awaiting the next tidal surge to cover them. Only then will they open, slowly sifting the water for food. A lone light-gray shape seems at first only an upturned log, exposed by the lowering tide. On slender legs it stands, and slowly moves its bent head, eyes peering at the mud. It is unusual for a night heron to be searching for prey in the bright sunlight, but a quick stab at an unwary fiddler crab attests to its foraging prowess.

Overhead a dark-gray herring gull glides to a distant shore, lands, and begins to slowly tear apart a dead fish. Soon he is joined by another, and forced to leave his meal to the more aggressive adult herring gull. The young gull flies in search of other carrion lying on the exposed mud. The nearby marsh grass sways gently in the breeze, and a myriad of blue-greens and yellow-greens form a dazzling carpet.

Although the quiet scene is typical of East Coast marshes, this particular marsh is often ignored, discounted, and spoiled by our everyday activities. For this marsh, and others like it, fringes the Arthur Kill, the waterway that flows between New Jersey and Staten Island. Except for local naturalists and recreationists who examine the Kill for wildlife, dangle lines to hook crabs, or travel the Kill in small boats in search of fish, it is all but forgotten amidst oil refineries, factories, docks, and garbage barges. Until the recent rash of oil spills, barge explosions, and tanker mishaps, most of us chose to ignore the Kill, preferring to think of it as just another "roadway" necessary for oil and garbage barge traffic—certainly not as an important ecosystem worth protecting.

It is, however, a matter of viewpoint. How do we normally view this river? From whose perspective is the Arthur Kill evaluated, judged, prized, or ignored? Most of us view the Kill from solid land, from our roads, highways, and bridges. We look through a mass of steel, through wooden buildings, old piers, and electrical wires. The marshes are obscured, the river a distant and unappealing waterway polluted by floating garbage, globs of black oil, films of shimmering oil, or tankers and garbage barges. It is one of the busiest waterways in the world.

The view from the river is much different: Drifting down the Kill in a small boat changes the perspective. The Kill is wide, the shores distant, the waves lap quietly on mud flats exposed at low tide. In some places buildings and piers meet the water, but in others only Spartina and mud fringe the Kill. Here and there the black mud is broken by stately white egrets, standing absolutely motionless, peering at the shallow, receding water. Beneath the water schools of killifish dart forward, unaware of the predator lurking above. With lightning speed the sharp, yellow bill of the great egret stabs, retrieves a large killifish, and swallows; then the egret becomes a statue, awaiting the next school. After feeding for half an hour, it lifts effortlessly and flies back to its nesting colony on Shooters Island.

Three islands lie within the Arthur Kill, each supporting an increasing number of nesting herons, egrets, and ibises. These birds build stick nests in the trees in March and April, lay eggs in May, and their young hatch in late May and early June. Their colonies are a mass of squawking, squabbling young, competing for the fish their parents bring back in their crops. The fish go to the chick that calls the loudest and can lift its bill the highest, grabbing the food from its parents' bill. By late August, the surviving young are able to fly and find food on their own. All is quiet in the colonies, the young and adults spread out over the Kill in search of food, and then they disperse out over marshes a hundred miles from the colony. Eventually they migrate south, leaving the heronry and marshes quiet, devoid of their stately, silent shapes.

Drifting along, one can see the water's surface broken by a tiny black head, eyes peering around. Then it is gone. Though once virtually extirpated from the area because of pollution and dwindling food resources, diamondback terrapins are making their way back into the marshes and tidal creeks of the Kill. A small estuarine turtle, diamondback terrapins live in the water, and females usually come on land only to dig nests and lay their nine to twelve eggs in late June or early July. Though not optimal habitat, the dredge-spoil islands provide a substrate for turtle nests, and a female may spend 1 to 3 hours searching for a suitable site before excavating the hole with her hind feet, laying the eggs, and covering up the nest. Slowly she rocks back over the nest, scooping back sand with first one foot and then the other. When the nest hole is full, she continues, moving 2 or 3 feet from the nest, obliterating the exact nest site from predators.

Years ago the Kill was very polluted, but then the culprit was raw sewage dumped in the river. Now it is unthinkable that we would dump in sewage. It's not unthinkable that we pollute it with oil, garbage, and other debris. But this too shall pass as our ecological awareness grows. The Arthur Kill ecosystem is by no means pristine. Many sensitive invertebrates and fish are gone. But perhaps not forever. The Kill is recovering from depleted oxygen levels and wanton dumping of toxins. It is a tidal system. And however slow the flushing is, it is still flushed out by tidal waters. It is connected to other marsh ecosystems that can provide colonizing populations—when the conditions are right.

From the viewpoint of our concrete byways, however, the Kill is a mass of piers, buildings, refineries, storage tanks, and chemical plants. Streets are narrow crowded by buildings. Somehow these roads are the last to be repaired, and large ruts and potholes are edged with jagged concrete and stone. Old chain-link fences protect the public from the once-toxic wasteland of Chemical Control Inc. (that repository of barrels of chemicals that went up in flames years ago, alerting us all to the dangers of toxic wastes).

In some places the buildings, factories and homes give way to small patches of green. But in the nutrient-poor soils only *Phragmites* grows, the tall, reedlike grass that invades many degraded habitats. It replaces our delicate, native cattails and rushes that normally grow in freshwater seeps and marshes.

The Kill is merely one example of many of our nation's, and indeed the world's, urban marshes and estuaries. Everywhere we must begin to inventory the organisms that live in these areas and improve their habitat, thereby improving our own. It is remarkable that these ecosystems can exist so close to our urban centers; they surely are a valuable resource worth preserving.

The Kill is a good example of how perspective or viewpoint affects our conservation, management, and development decisions. Because we see it from the land, from our wasteland of wires, wharves, and factories, we discount its wildlife and habitat value. This in itself condemns it to further degradation. Instead, I submit we should view it from the perspective of the organisms that live there. Where killifish school and spawn, diamondback terrapin nest, fiddler crabs scurry along the tidal creeks, and heronries exist, the marshes cannot be degraded beyond hope. The herons and egrets, and even the terrapin, may move elsewhere when disaster strikes, but the fish and fiddler crabs live there all year. And each spring the herons and egrets migrate back, to feed in the tidal marshes of Arthur Kill. It is our responsibility to protect their perspective.

Index

Achromobacter, 101
Acientobacter, 101
act of God, 47, 57; CERCLA definition, 47
advisories on fishing or crabbing, 6, 258–261
advisory group, 38–40, 71
Aequipecten irradians, 146
aesthetic injury, 49–50, 254, 261–262, 288–293
age effects, on fish, 188–194
air quality, legal injury to, 50
alewife, 9
algae, 6–7, 180, 182
alkanes, 37, 131
Alosa pseudoharengus, 9
American black duck, 218
American Littoral Society, 67, 196
American oyster, 146
ammonia, 5
Amoco Cadiz [ship], 147–148, 209, 265
amphipods, 8
anaerobic conditions, 5
analyses, chemical, 35–38
Anas platyrhynchos, 208, 210–214, 218, 227–234
Anas rubripes, 218
Anas strepera, 208, 218, 227–234
Anchoa mitchilli, 9, 178
anchovy, 9, 178
anemone, 180
anorexia, 94
Appleton, A. F., 64, 80
aromatic hydrocarbons, 101, 144, 149, 217, 239
Arthrobacter, 101
Arthur Kill, 2–10; biological features, 6–10; ecosystem, 2, 4–10, 15; physical features, 4–6
aspergillosis, 94–96
Aspergillus fumigatus, 94–96
assessment of spills: ecological monitoring, 15, 76, 160–163, 265–267; initial, 25–27
ataxia, 94
Atlantic bay scallop, 146
Atlantic silverside: life cycle, 179–180; pollution effects, 179; seasonality, 9

Atlantic strike team, 10
Aythya spp., 218, 229
Aythya valisineria, 218, 229

bacteria, 5, 101
Baker, J. M., 131–132, 139, 140
Balaena mysticetus, 242
baleen whale, 239, 242–243
Balthic macoma clam, mortality of, 142
barges, load capacity of, 11
barnacles, 182
baseline data, 28, 41, 178–237
bay anchovy, 178
behavioral impairment: in birds, 125–127, 211, 224–234; in fish, 189–195; in invertebrates, 110–124, 145, 150–151, 163–184; in reptiles, 124–125
benthic invertebrates, 8. *See also* invertebrates
benthic organisms, damage assessment, 51
benzene, 46, 151–152
benzopyrenes, 149
Bhopal, 276
bioamplification, *see* biomagnification
bioconcentration, *see* biomagnification
biodegradation, *see* bioremediation
bioindicators, *see* indicator species
biological monitoring, *see* biomonitoring; indicator species; monitoring
biological sampling, 31–35
biomagnification, 149, 216–217
biomass, changes after spills, 182–192
biomonitoring, 160–163; using birds, 232–234; using crabs, 163–174
bioremediation, 99–111, 181; factors affecting rate, 101–102; as a food production technology, 100; history, 99–102
birds, 51, 125–126, 222–229, 286; differential response of species, 83, 224–227; external effects of oil, 83, 126; fate of oiled birds, 34–35, 91–97, 218; foraging behavior, 226–227; hatchability, 83, 211, 224–225; inhibition of ovarian function, 83; internal effects of oil, 83; mortality, 31–35; nestling mortality, 225; reproductive effects, 83–86, 217–226; rescue and

birds (*continued*)
  rehabilitation, 68–71; treatment of
    oiled, 83–91
bird watching, 49, 253–256
bivalve mollusks, 142–159; total petroleum hydrocarbon levels of, 36
black-backed gull, *see* great black-backed
  gull
black-crowned night heron, 218, 223–229
black duck, 218
blood worm, mortality of, 142
blowout, 240, 245
blue crab, 142–159, 285; effects on
  and growth of, 150–152; human use of,
  253–266
bluefish, 9
blue mussel, 31, 149
boating, 254–255
Bonaparte's gull, 218
bottlenose dolphin, 243
bowhead whale, 242
Boy Scouts, 69
Branta canadensis, 90, 208, 211, 218,
  229–231
Brevibacterium, 101
brown algae, 182
Brzorad, J., 35, 65, 80, 140, 234, 257, 262
BT Nautilus [ship], 28, 57
Bubulcus ibis, 220–227
Bucephalus albeola, 218
bufflehead, 218
bulkhead, 6, 53, 82, 256
Burger, J., 35, 234
burning of *Spartina,* 132
Butorides striatus, 218

cadmium, 5
Callinectes sapidus, 253–266
Canada goose, 90, 208, 211, 218, 229–230
cancer risks, 260
Candida, 101
canvasback, 218, 229
Capitella capitata, 180, 182, 191–192
carbon, 7
Caretta caretta, 239, 246–247
carnivores, 10
cattle egret, 222–227
cell membranes, effects of oil on, 181
CERCLA, 24–25, 45–49, 57–58, 248; lia-

bility standard, 57–58; Type-A, Type-B
  rule, 24–25, 48
cetaceans, 242–243. *See also* marine
  mammals
chain of custody, 79
charismatic macrofauna, 78
Cheesequake, 116–124
Chelonia mydas, 239, 246–247
chemical analyses, 35, 38
Chernobyl, 276
chlorophyll, 8
chromatograph, 35
chronic oil spills, 132, 210, 216, 267, 277,
  286
Cibro Savannah [ship], 28, 36
Ciconiiformes, 7, 222–231. *See also*
  egrets; herons; ibises
Circus cyaneus, 7–8, 269
citizens, 1, 3, 64–80
civil-consent order, 38–39
civil liability, 44–54
clamming, 6, 254. *See also* advisories
clams, 31, 142–149, 285
cleanup, 23, 26; bioremediation, 99–111,
  181; conservationist view, 71–73; equipment, 10–11; owners/operators responsibility, 45; timing, 10–11
Clean Water Act, 24, 44–45
Coast Guard, 10, 15, 67; bioremediation,
  99–111, 181; cleanup, 10–11, 47;
  cleanup supervision, 10–11, 45; oil response protocols, 79
commensalism, 6
community, 272–275; effects on fish, 189–
  196; effects on waders, 222–224
community planning, 277–278, 287–288
community structure of fish, 189–196
compensable damages, 51–52
competition, 6, 15, 130; effect on fish communities, 189–196
Comprehensive Environmental Response,
  Compensation and Liability Act, *see*
  CERCLA
conservation, role of, 64–81
consumption advisories, 5, 258–261
consumptive use of natural resources, 49,
  257–262
contaminants, 5, 8, 46, 50, 171–180, 257–
  259
coordination, 23–43

cordgrass, 6, 10, 73–74, 130–141, 285; immediate oil effects on, 116–118; role in preventing predation, 204. *See also* Spartina
cores, 29–30
cormorants, 35, 165–208, 217–218; mortality, 35, 218
corporations as stakeholders, 274–275
Corynebacterium, 101
crabbing, 254–263
Crassostrea virginica, 146
crude oil, 102–103, 144–145, 152, 216, 239, 243, 266
crustaceans, 9, 144, 149–159, 253–266. *See also* fiddler crabs; grass shrimp
cycloalkanes, 131
Cynoscion regalis, 9
cytochrome P-450, 243

damage assessment, 1, 26–63, 69, 70, 72, 253–266; defined under OPA, 59; funding for, 27, 72; human uses, 253–266; initial collection of wildlife in, 34–35, 91–93, 126, 218; role of field notes of volunteers in, 70; and time lag for decisions, 72
DDT, 5, 151
decisions, 23, 72
deformities in birds, 5, 216–217
dehydration in birds, 85
density-dependent effects on birds, 224
Department of Environmental Conservation (N.Y.), 26, 103, 269
Department of Health, Welfare, and Housing, 26
Department of Interior, role in damage assessment, 50–52; CERCLA responsibility, 48
Department of Parks and Recreation (N.Y.C.), 26, 75, 103
Dermochelys coriacea, 239, 246–247
detritus in food webs, 7, 10
developmental defects, 5, 151–152, 216–217
diamondback terrapin, immediate effects on, 124–125
diatoms, 180
diesel fuel composition, 181
diet shifts, oil effects on, 226–227
dimethyl nitrosoamine, 152

dispersant treatments, 132, 210
dissolved oxygen, 5, 9
distance effects, on fiddler crabs, 171–172
dolphin, 238–239, 242–243
dose-response analysis, and health risk assessment, 160
double-crested cormorant, 165, 208–209; effects on in Arthur Kill, 217–218
dredge spoil, 7
ducks, mortality, 35. *See also* waterfowl
Dugongidae, 244–245
dugong, 244–245

eagle, 83
ecdysis, 151–152
ecological endpoints, 269–270
ecological risk, 160–162, 265–280
ecological risk assessment, 160–162; of the Arthur Kill oil spill, 269–272
economic analysis, human use of Kill, 253–266
ecotoxicology, 161
egret, effect of Arthur Kill spill on, 222–227. *See also* herons
Egretta thula, 222–229, 286
eider, 210
Eklof [ship], 28
Elizabeth, 23–26, 38, 51, 54
Elizabeth River, 4, 29
EMAP, 162
embryonic effects of oil on birds, 210
emergence, 116–125
endangered species, 79, 165, 269
Endangered Species Act, 248
endpoints, 269–270
Enhydra lutris, 243–244
enteritis, 85, 244
Enteromorpha spp., 182
Environmental and Occupational Health Sciences Institute (EOHSI), 174, 197, 278
environmental economics, 50
environmental monitoring, 160–174, 232
Environmental Protection Agency (EPA): bioremediation, 103; promulgating role under OPA, 59
equity issues, 39, 256–257, 261–262
Eretmochelys imbricata, 239, 246–247
Eschrichyitus robustus, 243
Eubalaena glaciilis, 242

European sea otter, 244
Exclusive Economic Zone, 44
existence values, 50. *See also* valuation; values
exposure assessment, health risk assessment, 160
Exxon: Arthur Kill spill, 1–2, 15; Bayway refinery, 10–11; and cleanup process, 45–46; compliance with "pay me" directive, 27; and cost of study, 27, 44; establishment of TPHC identity, 35–38; Oil Spill Advisory Committee, 71; oil spill details, 10–15; problems with human use study, 253–266; settlement, 1, 38–39, 44, 52–54, 73–75; wildlife rescue effort, 91–97
Exxon Valdez (ship), 51, 57, 101, 103, 115, 209, 215, 238, 244, 248, 262, 265, 267–268

feathers, 87–90, 206, 216; removing oil from, 87–90, 206
feeding, effects of oil on, 240–242
fertilization in bioremediation, 101–109
fiddler crab, 7, 117–124, 150, 160–177, 285; as bioindicators, 163–164; effects of oil on, 123–124, 160–177; immediate effects on, 117–124; mortality of, 164–165; reproductive problems of, 164; timing of exposure of, 150
fingerprinting of oil, 28, 35–38
Firth of Forth, 209
fish: biodiversity in Arthur Kill, 10, 178–179; community structure, 189–196; effects of pollution on, 178–182; sampling methods, 178, 183; species diversity of, 9; wintering behavior of, 191–192
fishing, 253–263
fishing advisories, 6, 258–261
flat oyster, 147
Flavobacterium, 101
Food and Drug Administration, tolerance limits, 50
food chain or food web, 5, 10, 117–128, 189–196, 222–224
food for human use, 257–266
foraging: in birds, 208–209, 211–212; effects of Arthur Kill spill in birds', 208, 212, 225–229
Forster's tern, 207
Fucus spp., 182

funding for damage assessment, 27, 38–39, 44, 52–54, 73–75
Fundulus heteroclitus, 9, 149, 178–179
fungal infection, following Arthur Kill oil spill, 11, 94–96
fungi in bioremediation, 101
fur seal, effects of oil on, 240–242

gadwall, 208
gasoline, 243, 266
geologic periods, 4
Geukensia demissus, 7, 31–33, 117, 120, 142–159
glossy ibis, 209, 223–229
Glycera americana, 142
government: conservationist view of role, 71–73; problems with rehabilitation, 96; role in settlement, 1, 38–39, 44, 52–54, 73–75; as a stakeholder 274–275
Governments Technical Committee (GTC), 26–29, 38
Grampus griseus, 243
grass shrimp, 9, 180, 183–196, 286; effect on from Arthur Kill spill, 183–196; life cycle, 180; size differences, 188–190
gray whale, 243; and grazing, food webs, 10
great black-backed gull, 201–222; effect on from Arthur Kill spill, 217–222; nesting on Pralls Island, 201–214
green algae, 182
green-backed heron, 218
green turtle, 239, 246–247
grey seal, 241
ground water, legal injury to, 50
grubby sculpin, 9
GTC, 26–29, 38
Gulf of Mexico, 246–247
gulls, 7, 35, 165, 201–222; effect on from Arthur Kill spill, 217–222; mortality, 35, 218; nesting shifts due to oil, 201–222; populations in Arthur Kill, 201–214

Hackensack River, 4, 29, 227, 283
hake, 10
Halichoerus grypus, 241
Harbor Herons Project, 64–65, 71–77, 80, 213, 234; *Harbor Herons Report*, 71, 75–77; role in cleanup, 71–73; role in settlement, 75–77

harbor seal, 241
hard-shelled clam, 31
hatchability: effect of oil on birds, 83, 211, 224–225; effect of oil on sea turtles, 247
hatching, 9
hawks, 8
hawksbill turtle, 239, 246–247
hazard identification in health risk assessment, 160
hazard: CERCLA definition of, 46–48; CERCLA liability, 58; laws governing, 24
hazardous wastes, CERCLA definition, 46
health advisories, 5, 6, 258–261
health risk, 6, 160, 259–261, 272
heavy metals, 5, 46, 149, 171, 180, 257
herbivores, 10, 130
herbivory, 130
heronries, 64, 222–227
herons: effects of Arthur Kill spill on, 222–227; mortality, 35, 218; reproductive success, 64, 222–227. See also waders
herring gull: effect of Arthur Kill spill on, 217–221; mortality, 218; nesting on Pralls Island, 201–214
hibernation: fiddler crab, 117–124; diamondback terrapin, 124–125
hiking, 49, 254
histopathological lesions: and oysters, 147; and soft-shelled clams, 146–149
hogchoker, 178
human health, 6, 259–261, 272
human use of Kills, 253–266
humpback whale, 242
hurricanes, 7
Hydranassa tricolor, 210
hyperplasia, and soft-shelled clams, 146
hypothermia, 244

ibis, effect of Arthur Kill spill on, 222–227
immersion time, 6
indicator species, 161–163, 269–270; birds, 232–234; crabs, 163–164
industrial discharges, 4
infection, in birds, 84
ingestion of oil by marine mammals, 241–244
inhalation of oil by marine mammals, 241–244
injury: assessment, 26–27; of birds, 34–35,

82–98, 125–127, 201–237; of blue crab, 150–151; of clams, 142–150; compensable, 51; defined legally, 46, 49; of diamondback terrapin, 124–125; of fiddler crab, 117–124, 150, 160–178; of fish, 178–200; general assessment methods, 160–163; of humans, 253–280; of mammals, 126, 239–252; proof of, 49–50; of resources, 24–25; of ribbed mussel, 32–33, 117–120, 142; of shrimp, 178–200; of vegetation, 116–118, 132–140
interagency cooperation, 23–43
International Geosphere Biosphere Program, 161–162
International Tanker Owners Pollution Federation, 77
interspecific variation in response, 189–193, 211
intertidal zone, 7
interviews, on human use of Arthur Kill, 257–266
invertebrates, 8, 30–39, 117–124, 142–200; general effects of oiling, 144–146; immediate effects, 117–124; initial sampling, 72; mortality, 31–33, 117–122
Isle of Meadows, 6, 10, 35, 223–227
Ixtoc I [ship], 219, 238, 246–247

Jamaica Bay, 3, 64
jellyfish, 8
jogging, 254–255
joint and several liability, 47
jurisdiction, 23–43

Kemp's ridley, 239, 246–247
killifish, response to mercury and dioxin, 149. See also mummichog
Kill van Kull, 4, 23, 28, 283

Lagenorhynchus acutus, 243
land acquisition, 75
land use planning, 64–65
Larus, 165
Larus argentatus, 201–214, 218
Larus atricilla, 207–221
Larus delawarensis, 218
Larus marinus, 201–214
Larus philadelphia, 218
larval stage, 9
laughing gull, 207–210

laws: table of, 45; oil spill, 44–63, 75, 248
lawyer's perspective, 1, 44–62
LC50 in toxicology, 144
Leach's storm petrel, 217
lead, 5
leatherback turtle, 239, 246–247
legal issues: civil liability, 44–52; context for spill, 24–25; criminal negligence, 53; juridiction, 52; lost use, 49; *Ohio* decision, 49–51; oil spill liability, 57; proof, 48; rebuttable presumption, 48; settlement, 53–57
legal settlement, 38–39, 44–63, 73–75
Lepidochelys kempi, 239, 246–247
liability: civil, 44–52; criminal, 54–57
life cycles: of fish, 178–180; of shrimp, 180
Littorina littorina, 142
loggerhead turtle, 239, 246–247
lost use, 49, 253–263
Louis Berger & Associates, 27–28, 30–43, 118, 120, 174, 218, 220; study design and results, 28–43, 218
lungworm, 241
Lutra lutra, 244

Macoma balthica, 142
macroinvertebrates, 10
Main Creek, 6
Malaclemys terrapin, 124–125
mallard, 208; mortality, 35, 218
mammals, 238–252; and oil exposure, 240
manatee, 238–239, 244–245
mangroves, 131
Manomet Bird Observatory, 35, 64–68, 231, 234; initial response, 66; role with TPL, 64–66
mapping, 29; of extent of oil spill, 12, 29–30, 36–38, 73; legal needs, 51; for rescue effort, 69–70
marinas, 255–256
Marine Mammal Protection Act, 248
marine mammals, 238–252
Maritrans [ship], 28
marshes, *see* salt marshes
marsh hawk, 7–8, 269
media coverage, 3, 65–67, 115, 130
Megaptera novaenglia, 242
meiofauna, 10

Melampus bidentatus, 7, 142
Menidia menidia, 9, 179–196
Mercenaria mercenaria, 31
mercury, 5, 149
Merrills Creek, 29
metals, *see* heavy metals; lead; mercury
microbial action in bioremediation, 99–102
Micrococcus, 101
microfauna, 10
migration: of birds and oil spill, 116, 126, 218–219; of fish and oil spill, 11, 191–192; vertical, 9
mixed-function oxidases, 149, 181, 241, 243
Modiolus demissus, 7, 31–33, 117, 120, 142–159
mollusks, 10. *See also* clams; mussels, oysters
molting in crustaceans, 151–152
monetary damages, 24–25
monetary settlement, 38–39, 53–59, 73–74
monitoring, 160–163; bioindicators for, 161; oil spills, 15; using birds, 232–234; using crabs, 163–174. *See also* indicator species
Morses Creek, 29
mortality: birds, 31–35, 210–211, 218–220; diamondback terrapin, 125–126; fiddler crab, 117–124; fish, 189–196; gulls, 218–220; invertebrates, 31–32; marine mammals, 240–245; mussels, 32–33, 117, 120; sea turtles, 245–246
mud crab, 9
mud snail, mortality, 142
mummichog, 285–286; biomass, 186–188; effects on from Arthur Kill spill, 183–196; life cycle, 178–179; pollution effects on, 149, 181; seasonality, 9; size differences, 185–187
muskrat, 126
mussels, 7, 117, 120, 142; mortality, 31, 142. *See also* ribbed mussel
mussel watch, 162–163
Mustelid, 238, 243–244
Mya arenaria, 31
Mysticetes, 142, 242–243
Mytilus edulis, 31
Myxocephalus aeneus, 9

naphthalene, 144, 151, 266
National Audubon Society, 64
National Oceanic and Atmospheric Admin-
    istration (NOAA), 25, 59, 103
native Americans, 24
natural areas, 3
natural resource: assessment, 24, 43–51;
    consumptive and non-consumptive, 49;
    damage to, 24, 43–63; OPA expansion of
    damages recovered, 58; trustees, 24,
    38–39
Natural Resources Defense Council, 11–
    12, 67
Neck Creek, 6, 29
necrosis, invertebrates, 147
nematode worms, 8, 180
nesting: effect of oil on beaches for, 247; of
    waders, 224–226
nest site selection, gulls and waterfowl,
    201–214
networking, 64–71
Newark Bay, 4, 8, 23, 31, 132, 151, 227,
    253, 283, 288
New Bedford Harbor, 271
New Jersey: Department of Environmen-
    tal Protection and Energy (NJDEPE),
    25–26, 73, 140, 174, 235, 269, 280; Divi-
    sion of Science and Research, 26, 29;
    marshes, 6–7, 29; Spill Act (see Spill
    Compensation and Control Act); spill
    compensation, initial assessment, 24,
    27, 52, 174, 197, 235
New York, 7, 29; marshes, 6
New York Bight, 8
New York City: Audubon Society, 64–71,
    234; Department of Environmental Pro-
    tection (NYDEP), 80, 284; Land Project
    role with TPL, 68–71; Parks and Recre-
    ation, 64, 103; refuges, 3
New York Harbor, 1
New York Times, 66
New York State: Department of Environ-
    mental Conservation (NYDEC), 39,
    103, 174, 269; Department of Health,
    258; Department of Law, 26
nitrogren, 7, 9, 101–103; effects on
    bioremediation, 101–103
NOAA, see National Oceanic and Atmo-
    spheric Administration
Nocardia, 101

northern harrier, 7–8, 269
No. 2 oil, 1, 10, 23, 28, 37–38, 68, 79, 91,
    106, 132, 142, 145, 150–151, 181, 210,
    266
nurseries, fish and shellfish, 7
nutrients, 7, 8, 101, 161; cycling, 7, 161;
    enrichment following bioremediation,
    101–102; load, 8
Nyctanassa violacea, 159, 269
Nycticorax nycticorax, 218, 222–229

Oceanodroma leucorhoa, 217
Odobenidae, 240–242
Odontocetes, 242–243
oil: chronic effects of, 146–147; composi-
    tion (see petroleum composition; individ-
    ual components); persistence of in
    saltmarsh peat, 138; removing from
    feathers, 87–90; tons released in world,
    216; weathering effects, 142, 144, 181,
    210–211, 239, 247. See also crude oil;
    No. 2 oil
oil effects: on birds, 31–35, 83–86, 91–97,
    205, 209–214, 216–220; on fish, 178–
    182; on invertebrates, 31–33, 117–124,
    142–146, 150, 160–177; on marine mam-
    mals, 239–240; on sea turtles, 238–240,
    246–248; on sediments, 29–30; on
    shrimp, 178–200; on terrapins, 125–126
Oil Pollution Act of 1990, 44, 57–59, 75,
    248
oil spill: conservation networking and, 64–
    71; description of Arthur Kill, 1, 10–15,
    283–284; extent of, 10–12, 35–40; first
    response to, 65–71; governmental re-
    sponse to, 67–68; human health effects
    of, 68–69; laws governing, 44–63; media
    response to, 3, 65–67, 115, 130; quan-
    tity, 23; rate, 15; and risk assessment,
    269–272; role of volunteers in, 68–71;
    and settlement, 53–59; weathering, 35–
    37
oil spills, 28, 266; chronic, 1, 132, 210,
    216, 267, 277, 286; future, 265–267; inci-
    dence 1982–1991, 13; major spills in
    N.Y. Harbor, 11, 266 (see also Amoco
    Cadiz; Exxon Valdez; Gulf of Mexico;
    Ixtoc; Persian Gulf); monitoring, 15; pro-
    tocol for study of, 76, 79, 288; open
    ocean rate, 15; rate in N.Y.–N.J.

oil spills (*continued*)
  Harbor, 12–13, 266–267; rates and
    amounts, 266–267; reducing risks of,
    276–277; response of conservationists
    to, 76
oil transfers, daily rate, 11
Old Place Creek, 6, 29
olefins, 131
Oncorhynchus gorbuscha, 181
Ondatra zibethicus, 126
OPA, *see* Oil Pollution Act of 1990
optimistic bias, 260
option value, 49. *See also* valuation; value
osprey, 83
Ostracods, 180
Ostrea edulis, 147
Ostrea lurida, 147
ovarian function, 83
oversight, 39
owner/operator, third-party responsibility,
    57–58
oxygen, 5, 131; effect on bioremediation,
    101–103

Pacific oyster, 147
Palaemonetes, 9
Pandion haliaetus, 83
parasitism, and marine mammals, 241
Parsons, K., 35, 213
pathologist, 79, 241–243
pathology, and marine mammals, 241–243
pay-me directive, 27
P-450 enzymes, 243
PCBs, 5–6, 46, 257; levels in crabs,
    258–259
pelage, effects of oil on, 239–240
periwinkle mortality, 142
Persian Gulf, 182, 215, 245; oil spill re-
    sponse in, 82, 113, 213, 245
pesticides, 46, 268
petroleum composition, 36–38, 149, 153,
    180, 217
Phalacrocorax auritus, 165, 208–209,
    217–218
pharmacokinetic models for invertebrates,
    150
Phoca hispida, 241
Phoca vitulina, 241
Phocidae effects of oil on, 240–242

phosphorus, 7, 9; effects on
    bioremediation, 101–103
Phragmites, 292
Physeter catadon, 243
physical contact, and marine mammals,
    240–243
physiological damage, 5. *See also* injury
Phytoplankton, 8
Piedmont lowlands, 4
Pierce, V., 94–96
Piles Creek, 29
Pine Barrens, 3
Pinelands Commission, 3
pink salmon, oil effects on, 181
pinnipeds, 238; effects of oil on, 240–242
plea bargain, 53
Plegadis falcinellus, 209
polar bear, 245–246
pollutants, 5, 8, 46; CERCLA definition
    of, 46
pollution effects on fish, 180–182
Polycheate worms, 8, 10, 180, 182
polychlorinated biphenyls, 5–6, 46,
    257–258
polydipsia, 246
Pomatomus saltatrix, 9
population: human, 2; oil effects on wader,
    223–224; pre- and post-spill levels of,
    219–220
Pralls Creek, fish sampling, 183–196
Pralls Island, 6, 7, 10, 64, 66, 215–237,
    284; bioremediation on, 99, 101–109;
    gull and waterfowl nesting on, 201–214;
    heron and egret nesting on, 222–227;
    initial wildlife rescue, 69
predation, 6, 15, 225; avoidance by young
    birds, 203; effect on fish communities,
    189–196
prediction of future spills in Arthur Kill,
    11, 265–267
preening, 206, 216
Presidente Rivera [ship], 57
primary productivity, 5–8, 10, 29, 161
Prince William Sound, *see Exxon Valdez*
productivity, 5–8, 29, 161; in gulls, 220–
    222; in waders, 226–229
property ownership, 29
protocol for preparedness, 76, 79, 288
protozoa, 180
Pseudomonas, 101

Pseudopleuronectes americanus, 178
public: risk perception, 3, 272–275; response, 11, 265–278; use, 253–266
Puffinus pacificus, 217

Rahway River, 4, 29; fish sampling, 183–196
Ramapo College, 72
Raritan Bay, 4–5, 8–9, 165–173, 227, 253, 255
recovery, 15, 270–272
recreation, 253–266, 272
recruitment of fish, 190–195
reference sample, extent of spill, 35–38
refinery locations, 14
regulations, oil pollution, 45. *See also* legal issues; Oil Pollution Act; Spill Compensation and Control Act
rehabilitation of wildlife, 1, 69, 73, 82–97; criteria for release, 90; definition, 82; removing oil from feathers, 87–90; response to large oil spills, 90–91; training required by settlement, 55
remediation: bioremediation, 99–111; burning, 132; dispersants, 132, 210
removal costs, 45
reproduction: birds, effects of oil on, 83; humans, effects of PCBs on, 259
reproductive effects, 5, 216–220; on gulls, 220–222; on waders, 222–227
reptiles, 73, 78, 124–125
resiliency, 270–272
resource assessment plan, 23–43; partitioning, 6–10
response: delayed, 23; tiered approach for Arthur Kill, 26
Rhodotorula, 101
ribbed mussel, 7, 31–33, 117, 120, 142–159; immediate effects of spill on, 117, 120; mortality, 31, 142
Richmond Creek, 6
right whale, 242
ring-billed gull, 218
ringed seal, 241
rip-rap, 6, 253
risk: definitions, 267–268; from eating fish and crabs, 258–261; ecological, 160, 265–278; estimations, biases that underestimate, 274–275; legal issues and, 55–56; for marine mammals from oil, 238–

239; reducing, 41; reducing from oil spills, 276–277
risk assessment: definitions, 265–268; hazard identification, 160
risk characterization, health risk assessment, 160
risk management, 267–268, 286
risk perception, 259–260, 286; and impact of oil spills, 275–276
Risso's dolphin, 243
rookeries, 64. *See also* heronries
rotifers, 180
Rutgers University, 35, 65, 68, 72, 100

Safe Harbor Coalition, 55–56, 71, 75, 80
salinity, 6, 15, 130, 191
salt hay, *see Spartina patens*
salt marshes, 6, 130–141; extent of oiling, 29, 134
sampling: biological, 28–40; sediment, 29
Sandwich tern, 210
Sawmill Creek, 6, 29
scaup, 218, 229
scavenging, secondary effects of oiling on, 69
scope of work, 25–27, 28–38
seabirds, 201–215, 217–222
sea lions, effects of oil on, 240–242
seals, *see* marine mammals; pinnipeds; sea lions
sea otters, 243–244
seasonal migrants: birds, 116, 126, 218–219; fish, 9, 191–192
sea turtles, 124–125, 238–252; route of oil exposure, 240, 246–247
secondary effects of oiling on scavengers, 69
Section 311, 44–45, 57–58
sediment, 29–31, 35–38; levels of TPHC, 36–40; sampling, 29–31
seines, use for fish sampling, 183–196
sensitive ecological areas, 41
settlement, legal 38–39, 44–63; initial, 1, 52–54; view from a conservationist, 72–73
sewage, 4–5, 8–11
shellfish, 6, 8; advisories (*see* advisories); beds, 24; shellfishing, 254–255. *See also* clams; mussels
ships, 8–11, 28, 36, 57, 147–148, 209, 269
Shooters Island, 10, 35, 217, 223–224

shoreline length impacted, 6, 29
shrimp, *see* grass shrimp
silverside: effects of Arthur Kill spill on, 9, 183–196; size differences, 188–190
sirenians, *see* dugong; manatee
slope, effect on nesting gulls of, 203–208
snails, 7, 8
snowy egret, 222–229, 286
soft-shelled clam, 31; gonadal development, 149; lesions, 146–149; mortality, 142; response to dioxins, 149
soil types, 4
Somateria mollisma, 210
Spartina, 6, 10, 285; and Arthur Kill spill, 132–140
Spartina alterniflora, 6, 73–74, 130–141
Spartina patens, 6, 130–141
spawning, 9
species diversity: birds, 7, 208, 215, 217–218; fish in the Arthur Kill, 9, 178–179; invertebrates, 7–9, 142–144
species sensitivity, 8
sperm whale, 243
Spill Compensation and Control Act of N.J., 24, 27, 39, 52, 174, 197, 235
spill response, maps of vulnerable natural resources, 67
sponges, 8
Sporobolomyces, 101
sports, 254–255
stakeholders, 272–275
Staten Island, 1, 2, 6, 10, 31, 253
statutes, oil pollution, 45
Sterna forsteri, 207
Sterna sandvicensis, 210
storage of oil: level in region, 11; locations, 14
storms, 7
surface feeders, 10
surface water, legal injury to, 50
susceptibility in birds, 210–212

tankers' load capacity, 8–11
temperature effects: on bioremediation, 101–109; on fiddler crab, 168–169; on waders, 223–227
temporal factors, human use of Arthur Kill, 260
thermoregulation, in marine mammals, 240–242

thin shells, 5
third-party defense, 57
tidal effects: on birds, 205–206, 225–227; on fiddler crab, 168–172; on fish and shrimp, 185–196; on foraging waders, 226
tidal flushing, 5
tidal gradient, 8
tidal range, 4
tidal waters, 4, 6
tissue sampling, 35–38
tolerance limits, injury, 50
toothed whales, 242–243
Torrey Canyon [ship], 182
total petroleum hydrocarbons analyses of levels (TPHC), 35–38; levels in sediments, 36–37
toxicity, acute effects of oil on invertebrates, 144–145, 239
TPL, role in cleanup, 64–73, 234
transfer facilities, 8–11, 14
Trichechidae, 244–245
Tricheus manatus, 244–245
tricolored heron, 210
Trinectes maculatus, 178
Tri-State Bird Rescue and Research, Inc., 68–70, 82–97
trophic level, 6, 10
trophic web, 6–10
Trust for Public Lands, 64–73, 234
trustee, 24, 38–39; CERCLA definition, 46–50; role under OPA, 59
turbidity, 9
Turisops truncatus, 243
turtles, *see* diamondback terrapin; sea turtles
type-A rule, 48–51
type-B rule, 48–51

Uca, 7, 117–124, 165–174
Uca pugnax, 165–174
Uca minax, 165–174
U.S. Dept. of Interior, 15, 25, 248; promulgating role under OPA, 59
U.S. Fish and Wildlife Service, 25
Urban Wilds Program, 64
Urophycis, 10
Ursids, 239
Ursus maritimus, 245–246

Valdez, see Exxon Valdez

valuation, 253–257, 261–262, 272–275, 290–293

value, 254–257, 261–262

values, 24, 49–50, 254–257, 261–262, 290–293; benefits, 24; market pricing, 50; non-use, 49; option, 49

vegetation, 130–141. *See also Spartina*

Vibrio, 101

volunteers 68, 73, 210; role in wildlife rescue, 68–71, 91–96

waders, effect of Arthur Kill spill on, 222–227

walking, 254–255

walrus, effects of oil on, 240–242

waterfowl: effects of Arthur Kill spill on, 227–234; populations in Arthur Kill, 210–214

waterproofing, birds, 83

water quality, 4–5, 162, 180; assessment, general methods, 162

weakfish, 9

weathering of oil, 35, 142, 144, 181, 211, 239, 247

wedge-tailed shearwater, 217

wetlands, 283

whales, 238–239, 242–243

white-sided dolphin, 243

wildlife rescue, 1, 68–69, 78, 82–97; post-spill treatment, 91–97

winter flounder, 178

wintering behavior community structure, 191–192

wintering effects, 113–127, 150, 153, 284; on fish and shrimp, 150–153, 190–194; of oil on gulls, 201, 217–222; of oil on waterfowl, 227–234

World Prodigy [ship], 57

worms, 8

wrack, effect on avian nesting, 204

yeast, 101

yellow-crowned night heron, 165, 269

zinc, 5

zooplankton, 8, 9